Praise for *The 5-Minute Gardener*

"*The 5-Minute Gardener* is the solution for anyone who dreams of cultivating a beautiful garden but feels short on time. As a busy entrepreneur and mom of three, time is not on my side, but I've had the opportunity to implement Nicole's advice in my own garden and watch it transform. With practical tips and easy-to-follow guidance, this book proves that you can nurture a thriving garden in just a few minutes a day."

— **Shea McGee,** *co-founder and CCO of Studio McGee and McGee & Co.*

"This book provides simple, accessible, inspiring gardening advice that any gardening nerd will enjoy. Gardeners at every level can benefit from Nicole's advice."

— *Jim Gaffigan, comedian, actor, writer, and producer*

"*The 5-Minute Gardener* empowers readers with the tools and information to cultivate a garden, grow our own food, and transform our relationship with nature— all things which have a profound impact on our health, our communities, and the planet."

— ***Casey Means, M.D.***, #1 New York Times *best-selling author of* Good Energy

"Starting a garden has been a dream come true for me and my family. Somewhere between growing my business and raising my girls, I added growing a tomato to my dream list—and finally checked it off! Nicole's simple, practical approach made it possible, and I can't recommend enough the power of just five minutes in the garden. If you're busy like me, this book will show you how to make gardening a part of your life."

— ***Jenna Kutcher,*** New York Times *best-selling author of* How Are You, Really? *and host of the* Goal Digger *podcast*

"I've been doing this long enough to know a true industry leader, and that's exactly who Nicole is. She combines innovation, real-world application, and a depth of caring for others, which has resulted in the garden revival sweeping the nation. *The 5-Minute Gardener* is a game-changer for anyone looking to bring more beauty, balance, and joy into their daily life."

— *Dean Graziosi, entrepreneur and* New York Times *best-selling author*

"If you want to start a garden, Nicole Johnsey Burke is the first person to call. And if you want it to thrive without losing your mind, she's the second person to call. We started and maintain a garden using her techniques and could not be happier!"

— *Donald Miller, author of* Building a StoryBrand

"As a business owner and a mom, I've always struggled with keeping up with managing a garden. Nicole's garden tips and teaching make it so easy to keep up with in a realistic and simplified way, and because of it, my garden is thriving!"

— *Jenna Rainey, artist and author of* Everyday Watercolor

"We often think we need to grow a garden, but what if it's the garden that grows us? Nicole transforms how we think about gardening, proving that even the busiest can cultivate a thriving garden—and, in turn, a flourishing life."

— *Lara Casey Isaacson, author of* Cultivate *and founder of* Cultivate What Matters

"Quick, easy, actionable tips in Nicole's signature style: coaching you through health and happiness with humor and humility. This book makes you want to garden. More importantly, it makes it all so doable, season by season!"

— *Bari Baumgardner, founder of SAGE Event Management*

the 5-Minute Gardener

ALSO BY
NICOLE JOHNSEY BURKE

*Leaves, Roots & Fruit: A Step-by-Step Guide
to Planting an Organic Kitchen Garden**

*Kitchen Garden Revival: A Modern Guide
to Creating a Stylish, Small-Scale,
Low-Maintenance Edible Garden*

*Available from Hay House

Please visit:

Hay House USA: www.hayhouse.com®
Hay House Australia: www.hayhouse.com.au
Hay House UK: www.hayhouse.co.uk
Hay House India: www.hayhouse.co.in

the 5-Minute Gardener

Year-Round Garden Habits for Busy People

Nicole Johnsey Burke

HAY HOUSE LLC

Carlsbad, California • New York City

London • Sydney • New Delhi

Published in the United States by: Hay House LLC: www.hayhouse.com®
• *Published in Australia by:* Hay House Australia Publishing Pty Ltd:
www.hayhouse.com.au • *Published in the United Kingdom by:* Hay
House UK Ltd: www.hayhouse.co.uk • *Published in India by:* Hay House
Publishers (India) Pvt Ltd: www.hayhouse.co.in

Cover design: Karla Schweer & Julie Davison
Interior design: Julie Davison
Interior illustrations: Sarah Simon
Indexer: Shapiro Indexing Services

The author of this book does not dispense medical advice or pre-
scribe the use of any technique as a form of treatment for physical, emotional, or medical problems without the advice of a physician, either
directly or indirectly. The intent of the author is only to offer information
of a general nature to help you in your quest for emotional, physical, and
spiritual well-being. In the event you use any of the information in this
book for yourself, the author and the publisher assume no responsibility
for your actions.

Cataloging-in-Publication Data is on file at the Library of Congress

Hardcover ISBN: 978-1-4019-7878-5
E-book ISBN: 978-1-4019-7879-2
Audiobook ISBN: 978-1-4019-7880-8

10 9 8 7 6 5 4 3 2 1
1st edition, January 2025

Printed in the United States of America

This product uses responsibly sourced papers and/or recycled materials.
For more information, see www.hayhouse.com.

For my children,
Carolyn, Brennan, Rebekah, and Elaine.
I write with the hope that you will
always live in a world full of gardeners.

CONTENTS

Preface

I didn't mean to laugh out loud.

"Mommy, when are you going to build me a garden?" It was my four-year-old talking while I was holding the baby and flipping (and burning) the grilled cheese.

"What was that?" I yelled over my shoulder, fanning the smoke alarm.

"My garden, when are you going to make it for me?"

She asked while I was in the middle of something. But for her, *this* was the something. She didn't care about the grilled cheese, the messy kitchen, or the errands we needed to run.

She knew it was summer, and she knew summers were for gardening.

What she didn't know was that there was barely time to brush my teeth, much less start a garden. With three small kids at home and another on the way, life was busy, if not chaotic. One step inside my living room and it was obvious: I wasn't thriving.

So when she asked when, not if, I'd start a garden for her, I had to laugh out loud.

But she had the gift of preschooler persistence. That wasn't the first time she'd asked, nor would it be the last. And perhaps because I secretly wanted a garden too, I decided this would be a power struggle that I'd "let" her win.

Long story short: she got her garden—well, she got some dirt and seeds and sunshine.

If you've read my first two books, you know that first garden was quite the disappointment. But my four-year-old never seemed to notice. Every morning after yogurt and cereal, every afternoon when she was the first up from her nap, and every evening as the sun started to set and the air cooled, she was out there, checking on her potatoes, measuring her sunflowers, and swinging back and forth from the deck, watching her garden grow right beneath her feet.

The timing was all wrong for starting that first garden. It was the middle of summer, we were renting, I was busy, and we knew so little. But at the same time, the timing was just right. We got our hands dirty, we dug in, and most of all, we learned valuable lessons that we would've had to learn the next summer if we'd waited.

What mattered that summer was not how beautiful or how productive the garden was. What mattered was that it happened.

We started. We tried. We waited. We learned. We smiled. We laughed. We promised we'd try again. And we kept that promise.

They say that you'll never *have* time for anything. If you want time, you must *make* it. And that's exactly what we did.

Looking back, I'm so grateful for that summer of bad timing. It changed the way I've spent nearly every moment since. That minute when I said yes to finding time to garden carved a new space in time for me that will always be reserved for the garden. Not because I'm less busy—if anything, life has only gotten faster—but because I'm more clear now on what the garden needs from me and what I need from the garden.

While technology and industry have their perks, they can lead to a disconnection with our food, with nature, and with the seasons that makes us think time is marked by a watch or a phone. When really, the place time counts most is outside, under the sky, where the garden grows.

If you've ever laughed out loud when someone suggested you take up gardening, or if you've dreamed of a life more connected to nature and the seasons and the food you eat but you can't seem to find the time in your everyday life to make it happen, or if you've ever used that four-letter word *B-U-S-Y* to describe yourself but you still would *L-O-V-E* some way to make a garden happen, this book is for you.

The 5-Minute Gardener isn't about homesteading, living off the land, or growing everything you eat. It's not about quitting your day job, becoming "crunchy," or whatever picture comes to mind that feels a little out of reach when it comes to gardening.

This book is for busy people, like me and you, who don't live to garden but garden to live. It's a guide that shows you how to fit the garden into your schedule in every season, each month, all 52 weeks, and even every day. It's gardening in 5-minute increments, for the fun and the delight and the gift of it.

It's gardening for the rest of us.

To make the most of your 5 minutes in the garden,
head to **fiveminutegardenerbook.com**,
where you'll find Gardener Habit Trackers,
journal prompts, daily tasks, and seasonal recipes.

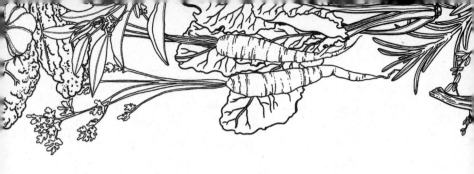

Introduction

*"All big things come
from small beginnings.
The seed of every habit is
a single, tiny decision."*

— JAMES CLEAR

When asked, most of us say we don't garden because we're not any good at it, that we're born with a brown thumb, that we're "too busy." But we know, deep down, that with time we could learn to grow at least a few plants.

The problem, then, is that in our current go-go-go lives, we just can't find the time to start learning or the endurance to keep a garden going.

Because what can you actually accomplish in 5 minutes?

More than you'd think. As I strive to build new habits into my life, I've noticed that the hardest ones to maintain are the ones I do on an irregular basis.

Tasks I do once a week or once a month become things I do once and never again. However, the tasks I do daily are the habits that stick: Going on a walk with my dogs every morning. Grabbing some greens for a morning smoothie. Making a cup of herbal tea and stretching before bed.

Tasks easily become part of my routine when I do them daily and when they take only a few minutes. Before long, these habits become second nature, such a big part of the way I spend my time that they start to define me as a person.

A 5-minute window may seem insignificant or "not enough," but it's actually just right if you want to be sure a new habit—like gardening!—sticks. Five minutes is enough time to walk through the garden, to deeply water a flower bed, to thin a patch of radish seedlings, to prune a tomato plant, to make a green smoothie, or to prepare a delicious dip with freshly harvested herbs.

If you wait to start gardening until you have more than 5 minutes, you may wait forever. But if you take action *in this minute*, for just a few minutes each day, you'll ultimately grow yourself into a gardener you'll hardly recognize.

*"I harvest greens for breakfast. After work,
I check on my plants and pull any weeds.
Then, after dark, I water the garden."*

— Megan

YOUR "WHY"

Most garden books teach the "how," but this book focuses on the "why" and the "when." Because once you define why you want to garden, you can then set goals and create habits to make the most of any 5-minute window you've got.

So . . . why do *you* want to garden?

Do you want to harvest a lot of great food? Spend meaningful time with your family or neighbors outside? Exercise or reduce your stress by surrounding yourself with plants?

Spend 5 minutes right now exploring your why.

For years, my "why" has been to use my garden as a center of health, the place where I go to get outside, get my steps in, create something beautiful, learn new skills, and appreciate the sky above me. As a bonus, I get to eat really good food. I garden because, deep down, I want to be as healthy and as centered and calm as possible. Because I have a lot of other goals beyond the garden: to change the world for the better, starting with myself, and then to help other people.

"[Gardening is] a postwork treat.
I go out while the kids are playing
and it's so therapeutic."

— Charlene

SET YOUR GOALS

Once you know why you're gardening, you can determine the "what" and "when" by choosing three main goals for the year and three for each season.

Your three seasonal goals can be anything, but I recommend that at least one of them be something about eating. You're growing food in this garden, after all.

Having a food-production goal works because goals that are easy to measure are often the ones that we tend to accomplish more often.

Not to mention, we're also more likely to accomplish the goals we can tie to everyday activities. BJ Fogg, author of *Tiny Habits*, calls these prompts. As he says, you can't form habits without prompts. For him, putting his feet on the floor first thing in the morning, going to the bathroom, or brushing his teeth are prompts to start new, healthy habits. Gardeners have at least three natural daily prompts, affectionately called *breakfast*, *lunch*, and *dinner*.

So before you start building 5-minute habit stacks, choose three gardening goals for the coming calendar year. These are my three goals for the year:

1. Eat from the garden every day. At least one thing I've grown should be on my plate or in my glass each day of the year. I know that if I eat more of the things I've grown, I'll be inclined to grow more things—and there's no downside to that. What I eat or drink from the garden changes with each season. But the fact that I enjoy a bite (or two or three) from the garden every day is a goal that motivates me and keeps me growing.

2. Make my world better. This broad goal covers smaller ones, like incorporating more native plants in my garden, improving my composting skills, creating less waste, and providing nectar and food for birds, butterflies, and ladybugs. It can stretch further to bringing a harvest to a neighbor, sharing a big garden meal with friends, donating food to a shelter, or showing a friend how to make the most of her space.

3. Enjoy the garden daily. This is something that a goal-setter like me can easily overlook! But

studies have shown that if we don't feel good
from doing our habits, we'll eventually quit,
no matter how good we believe the habit to be.
Feeling good in the garden is the thing that
keeps me gardening. So one of my goals is to
sit, walk, and eat inside the garden as much as
possible. This can include inviting friends to
come hang there, spending time with my kids,
husband, or puppies in this place, or simply
sitting down right in the middle of the path
and filling my mouth with cherry tomatoes.

With your yearlong goals in hand, break them into
plans for each season. These will (and should) change as
you progress through the months. As an example, here are
my hopes for the first season of the year, even while there is
chance of frost.

1. Eat homegrown greens every day, whether
    it's sprouts or microgreens grown on my
    windowsill, a fresh salad, or sautéed greens
    from the garden.

2. Prepare a daily garden drink (juice, smoothie,
    tea, or herb-infused water) to get me drinking
    more water and fill me up so I'm not tempted
    by soda and other drink alternatives.

3. Put cut flowers on the kitchen table every
    week. I want my garden to feature flowers and
    native plants for the bees and butterflies. I've
    accomplished this when I can make a weekly
    arrangement for my kitchen table.

Once you have yearly and seasonal goals, your only job
is to work toward them 5 minutes at a time.

HOW TO USE THIS BOOK

I've organized this book so that no matter what time of year you pick it up, you can quickly flip to the relevant page to see what your 5 minutes in the garden can look like today, based on the season, the month, the week, the day, or even the hour. When we divide our time like this, it makes gardening more interesting *and* simpler, because it takes the mystery out of "what to do" and "when to do it."

In a nutshell, this is how the system works.

Step 1: Look up *your* current season. Are you in the cold, cool, warm, hot, or second season? (See the next chapter for a full discussion of the seasons.)

Step 2: Look up the current month of the season. Each season is divided into 3 months (even though some may be a little shorter or longer). The first month of every season is focused on planting, the second on tending, and the third on harvesting.

As a gardener, you are only ever doing one of three things: planting, tending, or harvesting. If you have seeds or new plants to install, or there's digging involved, it's planting. If you need to prune, water, trellis, or protect, it's tending. And if you're picking, cutting, cooking, preparing, or eating, it's harvesting. Planting, tending, and harvesting tasks can (and most likely will) happen each day, but it's helpful to know the priority for any given month, week, or day.

Step 3: Look up the current week of the month. In our system, week 1 of the planting month is for planting big plants, week 2 is for planting seeds, week 3 is for tending, and week 4 is for harvesting. The tending month is divided into weekly focuses like watering, supporting, and pruning. There are weekly assignments for the harvesting month as well. In the first week, you'll harvest herbs and greens; in the second, you'll focus on root crops; and in the third and fourth, you'll turn your pruners to the larger fruit harvests.

Step 4: Find the current day of the week. You can assign each day with a key task, making sure that the day you have the most time available is the day you spend on that month's focus. So, for example, day 1 is usually for planning the week, days 2 and 3 for planting, day 4 for watering, day 5 for feeding, day 6 for harvesting, and day 7 for cooking. If you only have one day a week for the garden, just prioritize the task that's most important for that week and month.

Step 5: Look up the time of day. Each day is naturally broken into three parts—morning, noon, and evening—that can be assigned tasks, such as planting in the morning, tending at noon, and harvesting in the evening. You'll also find in each chapter an example of what an ideal day could look like. You'll build your 5-minute habits for the coming week from there. Reminder: I don't expect you to be in the garden every morning, noon, *and* night for this habit-stacking practice to be effective. (I'm certainly not!) Simply select one of those times of the day to be your "for certain" moment with your plants.

Then, when you have a few free moments, open to the page of this book for the current season, month, week, and even day. You'll immediately find a plan for how best to use your 5 minutes.

As you read through these suggestions, think about what time of the day would work best for you and commit to just one practice. You don't need to do all of the tasks mentioned or even most. But with a picture of the months and weeks of possibilities, it will be easier to imagine what a day could look for you, as a 5-minute gardener, based on the three tasks of a gardener: plant, tend, and harvest.

Remember, you only need 5 minutes—whenever you can find them.

One reason so many of us fail to develop a new habit is because we can't answer the question, *What should I do right now?* This book answers that question by showing you how to categorize your garden tasks and keep them simple so you never have to guess what to do next.

As a busy mom and business owner, trust me when I say that having a plan means I'm much more likely to go out to the garden whenever time allows. On any given day, I already have an idea of what's most important.

So if it's a Wednesday in the third week of July, I know I'm in the harvest month of my warm season, I'm in the tending week, and this is the day that I focus most on planting new seeds in vacant spots in the garden. With this plan, I can make the most of my minutes.

In another season, I might think, "Oh, it's the first month of the cool season, so I focus on planting. It's the second week; I'll focus on medium-size plants. It's the beginning of the week; I'll focus on planning the week ahead. And it's lunchtime, so I'll just step outside and observe what the weather is like today." For those of us who've never experienced gardening as an everyday routine, this book becomes our new calendar—a place to find what to do anytime we've got a moment to do it.

If you're ever lost or confused, or just need some clear direction, you can refer to the "Give Me Five" lists throughout the chapters as well as the "Quick Picks" list at the end of each season. There's always something you can do in a short amount of time that can make a big difference in your garden. Those minutes of progress add up faster than you think. The trick is to not waste a free second when you've got it, and this book makes that possible.

And for extra inspiration, turn to the Gardener Time quotes and tips throughout this book from other 5-minute gardeners. You'll also find recipes plus meal ideas under the

"Real Fast Food" headings—simple yet tasty dishes you can quickly pull together with your fresh-picked produce.

Now that you know how the 5-minute gardener system works, it's time to work the system. Begin by learning your seasons in the next chapter, and then start your first 5-minute practice right away.

> To make the most of your 5 minutes in the garden, head to **fiveminutegardenerbook.com**, where you'll find Gardener Habit Trackers, journal prompts, daily tasks, and seasonal recipes.

The Seasons

The garden keeps a pace and time that affects all of us, whether we recognize it or not.

Our eyes notice the bare trees and brown grass, the tiny pink buds, the waves of green, the bright orange and red leaves, and the seeds dropping under the trees.

Our skin feels the passage of time with the cool of the morning, the heat of the afternoon, and the heaviness of the evening.

Our nose smells the sweet buds in the morning air, the musk of hot nights, the freshness of fall, and the dry, crisp frost.

Our ears hear the tweets of birds building nests; the bees, flies, and hornets buzzing; the squawking of geese; and the silence of a frigid sky.

Our mouths taste time with the warmth of soup, the crispness of young lettuce, the sweet drips of juicy melon, and the soft texture of pumpkin pie.

We notice time every time we open our eyes or step outside.

If you're anything like me, when you open a gardening book or watch a gardener online, you try to follow their directions and do what they're doing, and then you realize they don't live where you live. They don't garden where you garden. They don't have the same weather that you do. Their seasons are different.

So before you can plan your year of 5 minutes in the garden, you need to know what *your* particular year looks like.

Perhaps you've heard of gardening zones or frost hardiness zones. You've seen a colored map on the back of seed packages or on plant tags and you wondered, *What does all that mean?* or *What zone am I?* or *Does it matter?*

The 5-minute gardener system goes beyond the zone so you can enjoy some aspect of the garden every day of the year. My seasonal system shows you what to do every day year-round, no matter where you are on a zone map.

The first step to understand your seasons is simply to find the average high and low temperatures in your town or city for each month. You can get more specific and record the highs and lows throughout each month, but that level of detail isn't necessary.

You can do this through a simple online search. Type in "average high and low temperature" in the search bar and then add the name of your city or town. Depending on where you live, the average high might be 95°F (or 85°, 75°, 65°, or 35°F) and the average low might be 20°F (or 0°, 35°, or 55°F).

Chart that information for each month from January to December.

With those numbers in hand, let's translate those average highs and lows into seasons.

The **cold season** is when the average high is at freezing or below (32°F or lower) and there's always a chance of frost.

The **cool season** is when temperatures increase and the average highs range from 35° to 65°F. Temperatures are rising daily, but there's still a threat of frost each night.

Once the threat of frost is over, the **warm season** begins. Average high temperatures range from 65° to 85°F.

Then there's the **hot season**. During this time, the average low is generally above 70°F and the average high is over 85°F.

Not every climate has a hot season, especially those closer to the polar regions. And not every climate has a cold season, especially those near the equator. But almost every region has a cool and a warm season.

At the beginning of the year, in the coldest period of the year, you're in the cold or the cool season, possibly the warm season if you're in a tropical location.

In the coming weeks and months, as your part of the planet gets closer to the sun, day length and temperatures increase, moving you into a warmer season. The cold season becomes the cool season, the cool season becomes the warm season.

At the summer solstice, the time when daylight hours hit their maximum, you progress to the warmest season—from cool season to warm and from warm season to hot. The garden is in its most rapid production during this season, which lasts until your side of the planet begins to distance itself from the sun once again.

At this point, you experience what I call the "second season." As temperatures decline and daylight hours decrease, you repeat the season you experienced just before your climate's peak. For hot climates, the second season is a warm season. For cold or mild climates, this second season is a cool season. The second season is mostly a mirror of the first. The first season had a continual rise in temperature and day length; this second season will have a continual drop in day length and temperature.

The first season's plants had to hurry and complete their production before things got too warm. And the second season's plants have to hurry and complete their production before things become too cold.

Finally, as your part of the planet loops to its farthest point from the sun, you experience the winter solstice, the time when day length is shortest and temperatures head toward their lowest.

For most climates, this will bring you to your coldest season. For hot climates, this is the cool or warm season, and for cold climates, this is the cold season.

Your seasons, no matter where you garden, create a perfect arc—what I call the "arc of the seasons"—starting at their coldest and gradually rising to their hottest, then mirroring the first part of the year as it proceeds through the second, until we're right back where we began, thus creating an arc.

The Seasons

If you live in a subtropical or tropical area, you'll experience a smaller arc, while those who live in polar regions experience a larger one. Following the arc of the seasons is critical to becoming a 5-minute gardener.

Once you've identified the seasons for your location, you have created the first part of your 5-minute gardener plan. You now know your seasons for the entire year, and next all you have to do is match your precious minutes with what tasks are needed on any given day of the season. (There's a section for each and every season in this book, so you'll always know how to use your time wisely, no matter what time of year you're reading this.)

Your garden schedule will arc alongside the seasons. You do small things in the cold and cool seasons, and the tasks get bigger as the days get longer and the garden grows, with the busiest and fullest time right there in the middle of the year. Then, at the end of the year, you'll repeat a lot of what you did at the beginning, like a mirror's reflection.

Now that you understand the arc of your seasons, it's time to match your gardening habits to where you are on this calendar. With the 5-minute system, you'll always know what to do in the precious minutes you have available each day, day after day, month after month, season after season, no matter where you're gardening.

So, now it's your choice. You can read this book beginning with the first season—the cold one. Or you can turn to the pages for the season you're currently in and start from there. You don't have to wait till spring or the first of the year, or some other "ideal" season, to get started. Begin in this moment.

There's no wrong time to start gardening, and this book will show you there's always something you can grow, 5 minutes at a time.

Year at a Glance

I've gardened year-round in three different climates, from Houston to Chicago. Here's what a year as a 5-minute gardener might look like in each spot.

A YEAR IN A HOT-CLIMATE GARDEN
Such as Houston

Cool Season (December–February)
December: Plant
January: Tend
February: Harvest

Warm Season (March–May)
March: Plant
April: Tend
May: Harvest

Hot Season (June–August)
June: Plant
July: Tend
August: Harvest

Second (Warm) Season (September–November)
September: Plant
October: Tend
November: Harvest

A YEAR IN A COLD-CLIMATE GARDEN
Such as Chicago

Cold Season (November–February)
November: Plant
December: Tend
January: Harvest

Cool Season (March–May)
March: Plant
April: Tend
May: Harvest

Warm Season (May–August)
May: Plant
June: Tend
July–August: Harvest

Second (Cool) Season (August–October)
August: Plant
September: Tend
October: Harvest

A YEAR IN A MILD-CLIMATE GARDEN
Such as Nashville

Cold Season (November–January)
November: Plant
December: Tend
January: Harvest

Cool Season (February–April)
February: Plant
March: Tend
April: Harvest

Warm Season (May–August)
May: Plant
June: Tend
July: Harvest

Second (Cool) Season (August–October)
August: Plant
September: Tend
October: Harvest

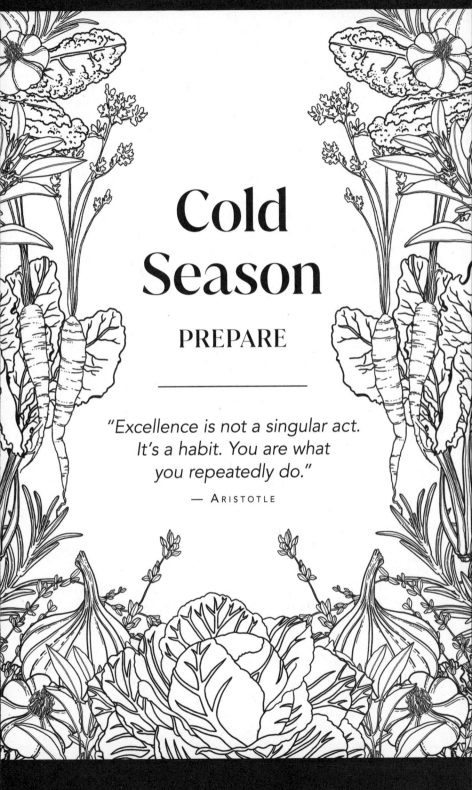

Cold Season

PREPARE

"Excellence is not a singular act. It's a habit. You are what you repeatedly do."

— Aristotle

If there's a season when you're the least likely to hear the word *gardening*, this is it. Instead, the word *cold* brings to mind images of rooftop snow, the fog of your breath in the outside air, and darkness that exceeds the hours of daylight.

It's true: the cold season is a time for rest. But more than that, it's a time to prepare.

The cold season is defined by daily high temperatures that barely pass the point of freezing, "short days" with just a few hours of sunlight overhead, and a consistent threat of snow, ice, sleet, or something frigid and wet that dares you to stay inside huddled by the fire.

Of course, not every climate has a cold season where temperatures stay below freezing for weeks and months. (If you don't have a cold season, you can still apply these practices for your coolest season, or you can skip to the Cool Season chapter and apply those habits to more of the calendar year.)

There isn't a lot of outside gardening in the cold season, but there is quite a bit to do inside and plenty of ways to use your 5 minutes to see green even if there's nothing but brown or white on the other side of the window.

The cold season is a time to enjoy simple treats and quick harvests—think fresh greens and sprouts, warm herbal teas, leafy green salads—and to cuddle up and prepare for the upcoming seasons when the sun will shine bright and long again. Your aim in this season is to make the most of the light and warmth you have while you enjoy the fruit of seasons past.

This is the perfect time to grow sprouts and microgreens—plants that grow with little to no sunlight and in minimal space and that can be harvested and eaten within a few days or weeks of planting. (See my book *Leaves, Roots & Fruit* for full details about winter growing.) The best part? Sprouts and microgreens are packed with nutrients and antioxidants in more concentrated forms than you can get from the full-grown produce at the grocery store and protect you from colds and flu in the season when you need it most.

It's also time to take the first steps to growing herbs and leafy greens and also start slower growing cool season and perennial plants indoors. By doing so, you'll get a jump-start on filling your garden, save loads of money when it comes time to plant outside, grow varieties you'd otherwise miss out on, and get a few small harvests you can enjoy even when the soil outside is frozen solid.

The first cold season month is set aside for planting. Plants and cold may seem like enemies but, in reality, the timing couldn't be more ideal. In this month, focus on getting as many seeds ordered and planted as possible in preparation for the cool season that's coming soon.

The second cold season month is for tending, taking care of the hard work you put in during month one, and ensuring everything you planted is showing signs of growth. There are only four steps of tending, and you can simply focus on one area each week.

The third cold season month is a time to finish up your indoor projects and fully enjoy all that's grown so far. Make huge sprout salads and microgreen dishes and cook up some pea shoots. Enjoy herb-infused sauces, breads, and soups.

Snow might still be on the ground outside, but if you follow the 5-minute plans, you'll be enjoying homegrown greens inside all season long.

*"I have young children, so often I will
welcome them into the garden with me.
If I need a break, I go to the garden, and
5 minutes easily can become 20 minutes."*

— Sarah

WHAT'S GROWING OUTSIDE IN THE COLD SEASON

While you're planting and growing indoors, loads of plants can be hiding under the snow and preparing to grow as soon as the sun comes out. The key is to directly plant these seeds in the soil before the ground freezes, so they'll be the first to sprout and grow when the weather warms.

Leaves	Roots
Cabbage	Beet
Collards	Carrot
Kale	Garlic
Mustard	Onion
Spinach	Radish
Swiss chard	

Cold Season at a Glance

MONTH 1: Planning the garden; planting sprouts, microgreens; starting cool season plants indoors

Week 1: Plan the garden.

Week 2: Start sprouts and microgreens.

Week 3: Start herbs indoors.

Week 4: Start large plants indoors.

MONTH 2: Tending sprouts and greens, indoor seedlings, outdoor seeds and plants

    Week 1: Feed plants.

    Week 2: Support seedlings and indoor herbs.

    Week 3: Prune plants.

    Week 4: Defend seedlings, indoor herbs, and microgreens.

MONTH 3: Harvesting cold season plants

    Week 1: Harvest sprouts and microgreens.

    Week 2: Harvest sprouts, microgreens, and herbs.

    Week 3: Harvest sprouts, microgreens, herbs, and small greens.

    Week 4: Harvest sprouts, microgreens, herbs, and small greens.

MONTHS

The months can seem to drag on forever in the coldest parts of the year, but having a new focus with each new moon can keep you going and help you believe that the sun really will come out tomorrow. In the first month of this season, you focus on planting; in the middle month, you focus on tending; and in the third month, you focus on harvesting. You may have previously assumed that in the middle of a cold season, *none* of those things would be happening. But you'll be surprised to see what's possible when you focus on gardening just 5 minutes at a time.

COLD SEASON/MONTH 1: PLANTING

Planting in the middle of winter? If you want to grow the best and healthiest plants with the biggest variety of colors and textures, get at least a month's head start on your first harvests, and save lots of money at the plant store, then the answer is yes.

Planting in the cold season doesn't just provide you a fuller, faster-producing garden; it also keeps you happy and healthy through the darkest and coldest part of the year. I've learned this after more than a decade of growing my own food. We simply cannot plant a seed without feeling at least a twinge of hope and expectation. It's impossible to dig a little hole, bury that tiny, seemingly lifeless stone, and not feel anticipation for the day it will peek its tiny head of green out of that dark and cold spot and remind us that life is indeed magical, at least in the garden.

So grab your dibber and some dirt and spend the first month of the coldest time of the year digging, planting, and hoping.

Cold Season/Month 1/Week 1

The first cold season month may be for planting, but you can't plant a thing until you've gathered the needed supplies and materials for growing sprouts, microgreens, perennial herbs, flowers, and large cool season plants. Use 5 minutes per day throughout the week to pick your varieties, choose your priorities, and double-check your list. Then place your seed order before the end of the week.

I recommend getting the highest-quality seeds you can afford, purchasing organic and heirloom variety seeds, if possible. You can find an alternative to almost every other garden tool and supply, but there's no substitute for the right seed.

What to buy? For sprouts, I suggest ordering large quantities of a few varieties of sprout seeds, such as alfalfa, arugula, broccoli, kale, radish, and pea. For microgreens, order bulk varieties of your favorite flavors, like buttercrunch, red lettuce, sunflower, arugula, broccoli, kale, radish, spinach, beet, and Swiss chard.

Depending on the size of your outdoor garden, your order could include seeds for herbs you'll start indoors now for planting outside next season, like rosemary, oregano, thyme, sage, marjoram, mint, and basil. You'll also want to order seeds for cool season herbs like parsley, dill, and cilantro. Don't forget to order seeds for large greens like kale, Swiss chard, and mustards; for large plants like broccoli, Brussels sprouts, cabbage, and cauliflower; for root crops like carrots, radishes, and beets; and for fruiting plants like peas, snow peas, and fava beans.

Finally, consider ordering flower seeds for both the cool and warm seasons (some of these need a big head start), including pansies, violas, calendula, cosmos, marigolds, zinnias, nasturtiums, and petunias.

Use this first week to also order supplies for planting, like containers, grow lights, soil mix, and possibly heat mats. For sprouts, all you'll need is one draining and one nondraining container. For microgreens, you'll need a draining tray, a nondraining tray, coconut coir (also called coconut fiber) or a soil mix, and a source of light (a fan is optional). For herbs, flowers, and plants, you'll need seed cell containers, a nondraining tray, a soil blend, grow lights, and a heat mat.

Give Me Five

Spend just 5 minutes a day over the first week selecting your favorite seeds and cold season supplies. Spend the first day looking over all your options and the second day prioritizing your favorites. On the third day, narrow your list, and on the fourth, place your order. Use the final days of the week to follow the same process for your tools and supplies. There you go: you've just gathered all the supplies you need for the entire season.

Cold Season/Month 1/Week 2

Now that you've got your supplies, it's time to put seeds into the soil.

Ironically, the easiest seeds to start are sprouts, and there's no soil necessary. Soak sprout seeds overnight, or for at least 12 hours, in a nondraining container. Then rinse the seeds thoroughly and place them in a draining container. Continue to rinse the sprouts every 12 hours until they are 4 to 6 inches tall and ready to be eaten. Cover sprouts in the first days of growth, then uncover to expose them to light in the final day so that you harvest delicious green sprouts full of nutrients.

Microgreens are nearly as simple to grow and will provide the flavors of summer even in the middle of the coldest winter. Begin by soaking coconut coir (or a preferred soil mix) overnight in a nondraining tray. Twelve hours later, spread coconut coir evenly in shallow draining trays. Sprinkle microgreens seeds on the coir. Tamp the seeds down and sprinkle less than 1 inch of coir on top of the seeds.

Place the draining tray into a nondraining tray and cover the tray with a simple cover, such as a cheesecloth, a dish towel, or a plate, to maintain humidity until the seeds sprout. At the first sign of growth, remove the cover and place the tray under grow lights set just a few inches above the leaves.

Later this week, start seeds for leafy greens, perennial herbs, and flowers. Use a seed-starting mix or make your own with coconut coir, earthworm castings, and compost. Soak the seed-starting mix until it's completely moist, then spread the mix into seed cells. Place one seed into each cell, being careful not to push seeds too deep into the soil mixture (only bury the seed two times its width). Place the draining seed trays onto nondraining trays and put a cover on top of the trays to maintain moisture. Once the seeds sprout, remove the cover and place the plants under full-spectrum grow lights.

Start an Indoor Herb Garden

Growing microgreens, sprouts, and leafy greens will give you hope during winter's chill, but having an indoor herb garden offers so much more—the delicious flavors of summer right in the middle of the coldest part of the year.

It takes just a few minutes to set up an indoor herb garden and less than 5 minutes a week to maintain and enjoy. Here's how:

Select the container. If you've got an extra-wide window ledge, an available table to scoot next to a window, or a little sunroom, consider creating an herb garden planter that combines several types of herbs (or different varieties of your favorite herb). Growing herbs together in a larger container is a little easier than using individual pots because the soil won't dry out as quickly.

The planter, pot, or container should be big enough to fit the herb's root ball. Choose a pot or container that's 12 inches deep and at least 12 inches wide so that you can grow several different types of herbs together.

When selecting your container, choose natural materials. My favorites are cedar, steel, and terra-cotta. Look for terms like *food grade* and *untreated* to ensure you're using the most natural materials for your edibles.

Ensure drainage. Make sure your pots have good drainage holes. Herbs hate to have their roots sitting in water. In fact, the surest way to kill an herb is to overwater it in a container with poor drainage. If your container doesn't already have good drainage holes in the bottom, make some with a drill. Space the holes every 3 to 4 inches.

Line the bottom. Cut a small piece of burlap to fit inside the bottom of the container to keep soil from running out of the pot and making a mess every time you water.

Add soil. Fill your container with a well-draining organic soil. You can use my favorite soil blend of one-third sand, one-third compost, and one-third topsoil. Top-dress with 2 to 3 inches of compost (my favorite is organic mushroom compost) to give your herbs a great start.

Plant. Select your favorite herbs from the list below. Place draping herbs near the sides so that they can cascade over the edges. And give young plants plenty of space to spread out and grow to their full potential.

Using this intensive planting method means you'll need to harvest leaves often to ensure each herb plant has access to sunlight and air circulation.

Place in the light. Most herbs only need 4 to 6 hours of sunlight a day, though they'll produce better if they're given more hours of light.

You'll know your herbs need additional light if they start getting leggy (tall and spindly) and/or the new leaves seem stunted compared to older ones. In that case, move the container closer to the window or set up a grow light. It's unlikely your herbs will receive too much light indoors, especially during winter, but if they do, you'll notice leaves that look bleached or scorched by the sun, or the plant wilts midday despite the soil being moist.

The Best Herbs for Indoor Growing

The best herbs to grow indoors include:

- Chives
- Marjoram
- Mint
- Sage
- Thyme
- Winter savory

Chives and other greens from the onion family may be the easiest to grow indoors. In fact, it was a small pot of chives—a gift from my mom—that taught me the simple pleasure of cutting something green to toss on our winter meals and got me into gardening in the first place.

"*Gardening is my therapy.*
When I'm stressed, I go out to water,
plant seeds, and admire all that's grown."

— Kristin

Cold Season/Month 1/Week 3

It was clearly January. My basement smelled like a forest and the floor was covered in dirt. Napa cabbage, kale, and Swiss chard were growing under the lights, and at least six trays of microgreens, spinach, and lettuce were stacked on the shelf.

I had more hope and plans than I had space, so I would rotate my trays into and out of my seed-starting shelves every morning and evening. I was still going to the gym, but you could argue that I didn't really need to—I was getting my steps in with all those trips up and down the basement stairs, and my arms were handling some serious weight with each and every tray.

The planting stage is both exciting and overwhelming. There's so much to do, and it's easy to run out of space, supplies, and energy. They say there's no one more out of touch with reality than the gardener starting seeds in January, and my winter routine is evidence that *they* know what they're talking about.

In the third week of the cold season, you'll continue planting any seeds that didn't arrive in time for the first round of planting. Each week, you'll start a new round of seeds for sprouts and microgreens to keep a continuous supply on your dinner table.

As you plant, prioritize the foods you want to grow most: the ones you can't find at the store and the plants that need the most time to take off and start thriving. For me, this includes leafy greens like napa cabbage, 'Blue Curled Scotch' kale, 'Toscano' kale, 'Bright Lights' Swiss chard, rosemary, oregano, thyme, and sage.

Over time, you'll discover which plants are the most important to you too.

Cold Season/Month 1/Week 4

By the fourth week, most of your seeds should be planted. And good news: if you included sprouts and microgreens in your planting month, you've already started harvesting.

This is a week to catch up and be certain everything is sprouting and growing. And it can also be a week for the first outdoor planting.

If you have a supply of clear containers like milk jugs, soda bottles, or clear storage bins, you can try winter sowing, an ingenious gardening technique that takes advantage of winter conditions to germinate seeds. By making mini greenhouses out of repurposed containers, you can get an early start to the growing season.

The idea behind winter sowing is that many seeds need to undergo a process called *stratification*, which involves exposure to cold and moist conditions, before they can germinate. By planting these seeds in containers that are placed outside and exposed to the ups and downs of winter weather, they undergo a natural stratification process. As temperatures warm in spring, the seeds germinate within their mini greenhouses.

Winter sowing of seeds is perfect for the 5-minute gardener. Simply slice the clear containers nearly in half widthwise, fill the bottom half with a well-draining potting mix, scatter seeds on the soil surface, and then securely close the containers (duct tape works well). Add several ventilation holes in the top and bottom of the containers and set the containers outside, even when it's freezing.

Little maintenance is needed once the seeds are planted. The transparent or semitransparent containers let in sunlight, while the ventilation holes allow for air and water too, creating a greenhouse effect that promotes germination when the conditions are right.

The best part? The seeds are less likely to be eaten by birds or rodents and are protected from major fluctuations in temperature. This method also reduces the need for hardening off the seedlings (that is, gradually exposing seedlings to outdoor conditions), as they gradually warm up to the outdoor conditions.

When the weather starts to warm, watch for the seedlings sprouting in each container. Once the seedlings are mature (they'll have a strong, thick main stem and dark green leaves and have undergone a few weeks of hardening off) and the threat of frost has passed, they can be transplanted into the garden. This method not only simplifies the process of starting seeds but also aligns the gardening activities with the natural rhythms of the seasons. It's an effective, low-cost, and low-effort way to start a garden that works even with limited space or resources.

Here are a few seeds you can sow in winter for growth in early spring.

Leaves
Broccoli
Brussels sprouts
Cabbage
Cauliflower
Chamomile
Cilantro
Dill
Kale
Lavender
Lettuce
Parsley
Spinach
Swiss chard

Flowers
Black-eyed Susan
Coneflower
Cornflower
Cosmos
Foxglove
Hollyhock
Larkspur
Lupine
Milkweed
Poppy
Sweet pea
Sweet William
Yarrow

Fruit
Snow peas and
   snap peas

Roots
Beets
Carrots
Garlic
Leeks
Onions
Radishes
Turnips

Additionally, local native flowers that are adapted to your region's climate are often excellent candidates for winter sowing.

When planting seeds, remember that not all seeds will germinate and become fully healthy plants. In the wild, plants hedge their bets by creating hundreds, if not thousands, of seeds each season in hopes that at least a few will make it from seed to fully grown plant. If nature knows to cover its bases, then you've got to do so as well: carefully sow a few more seeds than you need plants for the season to ensure that enough make it to maturity.

Create Your Own Seed-Starting Chart

Seed	Date Planted	Est. Date to Sprout	Date Sprouted	Date to Plant Outside	Est. First Harvest

COLD SEASON/MONTH 2: TENDING

Cold Season/Month 2/Week 1

In month 2, your key job is tending: to take care of the seeds and plants that are now 3 or 4 weeks old. These plants are changing daily and need your attention at each turn.

The tending tasks include feeding, supporting, pruning, and defending. These tending tasks become more complicated and take more time as the seasons progress. But in the cold season, most of the work can be done at the kitchen counter.

In the first week, be sure your plants have enough nutrients, water, and light. Plants "eat" differently than animals do, pulling nutrients through their roots and leaves, absorbing water, light, and the minerals they need to grow to the next level.

First comes water. Be sure you water just enough but not too much. Water is the medium your plants use to move nutrients and energy through their cells. So monitoring moisture is the first step to keep plants thriving.

A nondraining tray helps moderate the water your new seedlings receive each day. Fill the nondraining tray with an inch of water and observe how soon the plants absorb it into the soil. Use a water meter when you're unsure if your plants need water or are doing just fine.

Next, focus on light. Be sure all your plants receive an equal portion of light, turning the trays or adding more grow lights when necessary.

Finally, consider nutrients. The seed-starting mix used to begin the growing process has been supporting the growth of leaves and stems for a few weeks. So it's time to replenish the food source for your plants. You can apply a dose of fish fertilizer or earthworm castings.

Cold Season/Month 2/Week 2

Week 2 is the perfect time to focus on supporting plants. This is the time to be sure that every seedling is in a cell large enough for the plant to grow to its full maturity.

Measure the plant above the soil level; if it measures twice the size of the pot it's in, that's a sign to move this plant to a bigger vessel.

If seedlings are starting to fall over or bend, it's time to make adjustments. First, raise the grow light above your seedlings so their stems are encouraged to grow tall. Next, consider providing a fan that gently blows air on your seedlings to help strengthen them.

Give your seedlings a minute of your attention every morning and evening to be sure you don't miss a thing. These tiny plants may not look like much, but they're actually doubling in size overnight, growing new cells and stretching themselves to do the hardest thing: go from seed to plant. A tiny habit of simply running your hand over your seedlings or lowering the grow light can make a huge difference to the plants' long-term health.

Give Me Five

Spend just 5 minutes on your sprouts. Start a new tray of sprouts by soaking seeds in a non-draining tray. Give your established sprout tray a quick soak and place it in the sink to drain. Once you have a fully grown tray of sprouts, give the sprouts one final rinse and spin-dry in a salad spinner. Enjoy a quick sprout sandwich and then stack the newly soaked seeds under the draining tray. You've just eaten food you grew yourself and set yourself up for many more homegrown meals in under 5 minutes!

Cold Season/Month 2/Week 3

Week 3: time to prune. You may be wondering what there is to cut from if you're only growing small plants indoors. Pruning is simply pulling away what's unnecessary or no longer thriving in order to make room for what is. At this point, your pruning goal is to ensure the healthiest plants have all the space they need.

First, prune away any leaves that are turning yellow or pale, pale green or appear to be less than thriving. These leaves pull energy from the plant and can cause healthier leaves to suffer.

Next, it's time to thin. Thinning seedlings gives each seedling enough space in its cell to grow to its full potential. To thin, choose the seedling that appears strongest, with the thickest stem and broadest leaves, in that cell. Once you've found your winner, simply cut the other seedlings at the soil level. Voilà—you've just given the healthiest-looking seedling the best chance at surviving and growing all the way to harvest.

Cold Season/Month 2/Week 4

In the final week of tending, it's time to defend and protect your seedlings from any challenges they face. By this point, the seedlings have sprouted and are growing their first set of true leaves. You should be harvesting sprouts weekly and microgreens at least once every other week.

If any plants aren't thriving, it's time to troubleshoot, address the problems, and decide if you can keep going or need to start over.

If your sprouts aren't sprouting . . .

- Sprout seeds weren't rinsed often enough.

- Sprout seeds dried out.

- Sprout seeds may be expired.

- Sprouts didn't fully drain.

If your microgreens aren't germinating . . .

- Soil isn't warm enough.

- Soil isn't moist enough.

- Soil is too moist.

If your seeds haven't sprouted . . .

- Soil isn't moist enough.

- Soil isn't the right temperature.

- Soil is too moist.

- Seeds are old and no longer viable.

If your seedlings look leggy, meaning their stems are not thick and strong . . .

- Lights are not close enough to the seedlings.

If your seedlings appear yellow or light green instead of the anticipated dark green . . .

- Soil needs more nutrients—add nitrogen-rich fertilizer.

Give Me Five

Follow these quick tips to keep your indoor herbs happy and healthy.

- Monitor the soil moisture often. Allow the top 1 to 2 inches of the soil to dry out before watering.

- Consider watering herbs from the bottom to discourage fungus gnats on indoor plants. Place a saucer or flat-bottomed container underneath the herb pot and fill it with water. Let the roots soak up the moisture for about 15 minutes, then discard any remaining water. If you prefer to water from above, discard the water in the plant's saucer after 15 minutes, or water plants in your kitchen sink so excess water drains away.

- Make sure the herb plants consistently get 4 to 6 hours of sunlight.

- Rotate the pot every couple of days so that the leaves receive light evenly. Supplement with artificial light if needed (more about light on page 54).

- Harvest frequently to encourage the plants to produce more leaves.

COLD SEASON/MONTH 3: HARVESTING

Don't underestimate the need for a reward when it comes to starting or cementing new habits. We know toddlers, kids, and puppies are motivated by treats, but we often forget that we as adults need to be rewarded for our hard work too. That's what the third month of each season is all about—taking time to celebrate all the time you've put in and enjoying the small benefits of every tiny habit you've created. So grab your scissors and treat yourself so you can keep yourself growing into the next season.

Cold Season/Month 3/Week 1

In the first week, you begin with easy-to-harvest sprouts. Simply give your sprouts one final rinse with water, put them in a salad spinner to spin-dry, and then enjoy.

Try to harvest some sprouts every day this week and use them atop salads, soups, omelets, and toasts. In fact, there are loads of ways to enjoy sprouts in everyday meals. See some of the best sprout dish ideas in the cold season Weeks section.

Cold Season/Month 3/Week 2

By the second week of harvest month, it's time to focus on microgreens, if you haven't already started cutting from your tiny plants.

The ideal time to harvest microgreens is when the first true leaves have just fully developed, but before the second set of true leaves emerge. This is when the plants are most nutrient-dense and flavorful. Using a sharp, clean pair of scissors, cut just above the soil line or growing medium.

After harvesting, wash microgreen leaves gently to remove any soil residue and run the greens through a salad spinner. Microgreens will wilt and change consistency quickly, so eat them as soon as you can.

Enjoy microgreens as garnishes or in salads, sandwiches, and wraps. Blend mild-flavored microgreens into smoothies. Salad microgreens will be mild and grassy, brassica (kale, arugula, mustard) microgreens will be peppery, pea shoots will have a savory rich flavor, and basil will have a spicy flavor.

If you get the timing right, the reward is a fresh, nutritious, and flavorful addition to a variety of dishes, even in the midst of the coldest parts of the year.

Cold Season/Month 3/Week 3

It's now time to begin harvesting herbs like oregano, chives, thyme, and lemon balm from either your indoor herb garden or your outdoor space.

These first cuts of fresh herbs mark the beginning of the gardening season, even though there's still plenty of cold days ahead. These herbs are frost-hardy and often among the first to show new growth as daylight hours begin to lengthen and temperatures rise.

Oregano is one of the first herbs to sprout in late winter. Wait until the plant has several inches of new growth before harvesting to be sure the plant is well established and can withstand trimming. Snip the stems just above a set of leaves so the plant branches out and becomes bushier. The young, tender leaves of oregano are aromatic and flavorful, making them perfect for fresh use or drying.

Chives, with their delicate, onion-like flavor, are another early riser in the herb garden. They are incredibly hardy and can be harvested as soon as the new shoots are several inches tall. Cut the chives about an inch above the soil level to encourage new growth, harvesting no more than one-third of the plant at a time, allowing it to recover and continue producing throughout the season. Chives are best used fresh, adding a mild, oniony flavor to salads, soups, and other dishes.

Thyme, a versatile and robust herb, also begins to show new growth in late winter. Snip the top few inches of the branches, just above a leaf node, to encourage the plant to become fuller and more productive. Thyme smells so good and can be used either fresh or dried.

Lemon balm has a citrus scent and is one of the first greens popping up in my garden long before spring arrives. As with other herbs, cut just above a leaf pair to encourage new growth. The leaves of lemon balm are delicate, so they are best used fresh in salads, desserts, and drinks to retain their fragrant lemony essence. Dried lemon balm is great in winter teas.

These early-harvest herbs not only provide the first fresh flavors of the season, but they give you a tiny glimpse of all the things you'll enjoy from the garden in the seasons to come.

Cold Season/Month 3/Week 4

If you're growing the start of leafy greens indoors, you may be able to harvest some of your first leaves before these plants even move to the outdoor garden.

As soon as each plant has eight or more leaves, you can begin to harvest your first homegrown salads of the cold season. Cut the outer and lower leaves, being careful not to remove more than a third of the plants' leaves during a single harvest.

Wash the leaves carefully to remove any dirt, then give them a quick rinse and spin in a salad spinner. After washing, lay the greens out on a tea towel or dishcloth to completely dry.

For the best taste, eat these greens right away in a fresh salad. But if necessary, wrap the greens in a tea towel and refrigerate in a sealed storage container.

Levels of Frost Protection in the Garden

When you start seeds outdoors in the cold season, you'll need to provide some protection from frost and changes in temperatures. As your plants begin to grow, the weather will definitely fluctuate—warm on some days and cooling off drastically on others. There are at least four different ways you can protect your outdoor plants to be sure their growth isn't ruined by a sudden cold snap.

Fabric. This is a "level one" DIY frost protection for plants, ideal for those occasional cold snaps when the temperature dips below 32°F overnight. You simply grab old towels, sheets, and even bedspreads and drape them over the plants. Fabric traps warmth radiating from the ground and from the soil. Fabric covers also prevent any moisture in or on your plants from coming into direct contact with freezing air.

The key to effective DIY frost protection is to use materials that fully cover your garden space and reach the ground. Again, the idea is to create a warmer microclimate under the cover to trap heat from the ground.

Because these covers don't allow sunlight to reach the plants, they should be removed as soon as temperatures rise above freezing. Even if you expect frost more than one night in a row, remove your covers for a bit during the day to give the sun time to warm the soil again. This also releases any moisture that was trapped underneath the cover.

Frost cloth and frost blankets. Frost cloth, also called *garden fleece*, is a great option if you've just planted seeds or seedlings that need frost protection, or if you want to extend your growing season for perennial herbs and frost-tolerant plants like Swiss chard, beets, kale, spinach, carrots, lettuce, and cabbage.

These are similar to old sheets and towels in that they trap warmth from the soil underneath the fabric. But these commercial products actually let in a bit of air and sun. How much air and sunlight varies from product to product. Some frost cloth is tissue-thin to allow plants to breathe; others are medium weight. Agricultural 30 frost cloth, for example, provides about six degrees of protection (meaning if the air temperature is 30°F, the air underneath the frost cloth will be about 36°F) and allows about 70 percent of sunlight through.

Tips for using frost cloth:

- Most frost cloth arrives in a big roll. You can easily cut more than you need and double it by folding it in half for more protection.

- Install garden hoops across the planting row to support the frost cloth so that plants don't come into direct contact with cold cloth.

- Tuck in your plants like you would a child at bedtime—snugger than a bug in a rug! Air pockets around the edge of the frost cloth can let outside air in, which can lead to frost damage on your plants. Tuck in every corner, even the back of your garden (this can sometimes be a challenge if your garden beds are up against a fence or structure).

- Use landscaping pins or garden staples to push through the edges of the frost cloth to hold it in place. If your growing bed has a 4-inch-wide wooden top trim like that in a Gardenary-style raised bed, you can also use clamps or greenhouse clips to secure your frost cloth around the edges of the trim. Alternatively, weight frost cloth down using rocks or bricks in hard-to-reach places, like the back of the garden.

Plastic sheeting row covers and hoop houses. Unlike frost cloth, plastic sheeting does not let water or air in, but it does admit a varying amount of sunlight, depending on the thickness of the plastic.

Level three allows you to grow cold-tolerant veggies like carrots and spinach during periods of cold weather, but know that these crops will likely slow down their growth. I've seen gardeners in very cold parts of the country use these plastic tents to keep their plants alive in winter, not necessarily thriving and growing, but alive.

Snow easily slides off plastic sheeting instead of accumulating, and it also adds 5° to 15°F more protection than just frost cloth alone. The hoops are important for structure because you don't want the plastic to touch your plants at all. (Plastic can trap moisture on the plant and contribute to frost damage.) Remove this plastic layer as soon as temperatures rise so that your plants get fresh air.

Tips for using plastic sheeting for frost protection:

- First install garden hoops or PVC pipes at each end of your garden bed and then every couple of feet in the middle, depending on how long your bed is. Your goal is to slide the hoops or pipes deep into the soil of your beds and create a nice, tall arch over your plants.

- Cover the hoops with frost cloth and secure it in place.

- Cover the hoops with plastic sheeting and secure it in place.

Cold frames and mini greenhouses. Short of having a large greenhouse, the best protection you can offer your plants is with a semipermanent structure like a cold frame or mini greenhouse. This level is best for those of you with a true cold season. That means you have months where your average temperature doesn't reach above 32°F and there's always a chance for frost or snow.

A cold frame is a five-sided structure with sides and roof made of glass or plastic. The structure is placed on the trim around a raised bed or directly on the soil around crops, and the angled roof of the cold frame should face south to capture as much sunlight as possible. The plants inside a cold frame will receive sunlight and enjoy temperatures about 10°F warmer than the outside air. (Keep in mind that factors like the level of humidity inside the frame and the amount of sunlight it receives can impact the temperature.)

Another option for protecting individual plants from frost and snow would be garden cloches, which are basically bell-shaped mini greenhouses.

"I love starting with microgreens to add nutrition into my everyday schedule."

— Whitney

WEEKS

As you look at the weeks ahead in the cold season, focus on one thing you can do each day. Think of the beginning of the week as the time to plan and plant, the middle of the week as the time to tend what you've planted, and the end of the week to enjoy a little taste of your hard work. Plan and plant at the beginning, tend in the middle, and enjoy at the end.

With that strategy in mind, let's consider what a week could look like in the cold season.

Day 1

Spend time on the first day of the week planning out the other 6 days based on the weather forecasts. Will there be exceptionally low temperatures or the threat of snow? If so, make plans to cover the most sensitive plants. Will there be milder days? If yes, schedule time to plant seeds and herbs outdoors or do some winter sowing.

Once you know the forecast for the week, plan your indoor tasks, like making planting plans, ordering seeds, or starting seeds indoors for the coldest days.

With a few minutes of weather monitoring, you can protect your plants, maximize your garden's potential, and enjoy gardening activities in a way that works with whatever comes this season.

Make a Planting Plan

They say "fail to plan and you plan to fail." And while that's not 100 percent true in the garden (plants can still work their magic), you'll see everything in the garden go to the next level when you start making a plan for it.

The first step in creating a plan is to draw out the measurements of each garden bed. You can use a separate sheet of paper for each bed. Working in pencil (to allow for adjustments), draw the length and width of each bed to scale so you can determine how many plants can actually fit into each space. Be sure to include any physical elements that will impact the plants, like trellises, cold frames, or other structures inside the planting beds.

Next, create a plant key with large, medium, and small circles to denote the different sizes of plants. Work to scale: large circles should represent 1 square foot in the garden, medium circles should represent about ¼ square foot each, and the smallest circle should equate to about ¹⁄₁₂ square foot. Use a unique color for each type of plant within one size so you can know which plant goes where once it's time to plant.

Create the number of circles of each size to show how many different plants will be within that particular garden bed. For a cool season planting plan, large-circle plants could include broccoli, kale, mustard, collard, Brussels sprouts, and cauliflower. Medium circles would indicate Swiss chard, celery, fennel, cilantro, parsley, or dill. And small circles would represent buttercrunch, arugula, radishes, carrots, or spring mix lettuce.

Once the bed is drawn to scale and the key created, it's time to select which plants and sizes will go into each raised bed. Start with the large plants and center them down the middle of the bed (or the back of a border bed). Then plant medium plants to both sides of the large ones. End with the small plants on the outside lines of the garden space.

With this plan in hand, you're ready to order seeds, start seedlings, and plant out into the garden as soon as the soil starts to warm up.

Day 2

It was not even Valentine's Day in Chicago and *freezing* was not a strong enough word to explain the weather. But the sun broke through the winter cloud cover one afternoon, and I decided to test my luck at planting some lettuce seeds in the garden.

I asked my youngest to join me. The garden soil was still quite hard, but we had several bags of compost that we could pour on top.

Cold Season

We cleared the garden of debris, spread the compost on the bed, and started planting into the compost. Three rows of spinach went down the middle of the bed, followed by spring mix seeds to either side of the spinach. We finished with rows of radish along the borders of both sides. Seeds planted, we then covered the bed with hoops and frost cloth, gave each other a high five, and quickly ran inside to warm our fingers.

The few neighbors who dared to walk outdoors on this frigid day looked at us like we were mad. And maybe we were, but being stuck indoors when it's freezing cold can do that to you. We felt a little less of whatever winter madness we were suffering from after working in the crisp air, touching soil and seeds, and smelling the promise of early spring harvests.

While your focus on day 1 of the week is planning, the plan for day 2 is to head outdoors to attempt some outdoor seed sowing.

The middle of the week in the cold season is meant for stretching the season. This is a day to venture outdoors or to plant a few new things indoors.

As soon as the soil is workable, you should make your day 2 focus to plant directly into the soil. If not, then you can plant in jugs (see page 34) or cover the soil with a frost cloth so that it will warm up enough to plant in a few weeks' time.

Direct sowing seeds under frost cloth during warmer winter days only works for frost-tolerant vegetables like kale, cabbage, spinach, Swiss chard, and a few varieties of lettuce as well as flowers like pansies and sweet peas.

Based on germination schedule, plant cabbage, broccoli, and cauliflower on day 2 in the first week. Plant Swiss chard and spinach on day 2 in the second week, and plant kale, romaine, and iceberg on day 2 in the third week. On day 2 of the fourth week, replant any seedlings that didn't sprout or that you've determined you need more of.

The first step is to prep your garden beds by loosening the soil and adding any necessary amendments, like compost. (If the soil is frozen to the surface, you'll need to postpone adding amendments for another week or more.) Once the bed is ready, sow the seeds according to their specific depth and spacing requirements. After sowing, cover the area with a frost cloth. Use landscape pins to attach the frost cloth to the soil.

Frost cloth is *essential* in this process. It acts as a barrier against harsh conditions, like frost and strong winds, while allowing light, air, and moisture to reach the seeds. The effect is a microclimate more conducive to germination and early growth than the unprotected outdoor environment.

Once your winter beds are planted, monitor the weather each day to know if additional protection like thicker frost cloth or a layer of plastic is needed when a sudden deep freeze is forecast. On days when the weather is milder, lift the cloth during the warmest part of the day to give the seeds some direct sunlight and fresh air, which is beneficial for growth.

Starting seeds outdoors gets you ahead in many ways. First, you'll end up with a longer harvest period. Second, plants that start in cooler conditions have stronger root systems, so they're more resilient when temperatures begin to rise. Finally, the plants will be slower to bolt than those transplanted from indoor growing conditions. (Bolting is the process where plants rapidly grow a flower stalk and go to seed, typically in response to longer daylight hours and warmer weather.)

Sure, it may take a while for your fingers to warm up once you get inside, but the few minutes of planting outdoors each week will pay you back with months of harvests you just can't get if you don't put on your coat, step outside, and dig in.

Day 3

Start this day with a new set of sprouts. Rotate between alfalfa, broccoli, and radish seeds. Just before bed, simply toss a tablespoon of seeds into a tray and add some water. Cover the soaking seeds and tuck yourself into bed too.

In the morning, give the soaked seeds a rinse and let the seeds slowly drip dry over a draining tray. While at the sink, soak a small block of coconut coir, spread it on a draining tray, and plant it with a new set of microgreens seeds. Try basil, Swiss chard, cilantro, radish, or kale to give you a new flavor for the following week.

Vary the plant family you choose for sprouts and microgreens. If you select a brassica plant for sprouts, select a plant from the carrot or lettuce family for your microgreens. This ensures you'll be eating lots of flavors and a wide variety of nutrients in the coming week.

Once sprouts and microgreens are started, you can take a few minutes to start a new tray of large plants that will move outdoors as soon as the soil is workable.

Have another few minutes? Use the coconut coir remaining from your microgreens tray and toss some earthworm castings into the blend. Create some soil blocks and plant out a quick tray of large leafy greens like kale, cabbage, broccoli, and cauliflower. Then plant a tray of Swiss chard and spinach. Finally, assemble a tray of head lettuces like iceberg, romaine, and buttercrunch—lettuces that take longer to form a full head benefit from being started indoors.

These are the seeds you start inside:

Leaves	Roots
Broccoli	Garlic
Brussels sprouts	Leeks
Cabbage	Onions
Cauliflower	
Kale	**Fruit**
Mint	Fava beans
Oregano	Peas
Rosemary	Snow peas
Sage	
Thyme	

Day 4

Day 4, the first tending day of the week, is all about feeding plants. This is the time to add any extra nutrients to your microgreens soil, to add more water to the nondraining tray of sprouts, and to add water and nutrients to your indoor seedlings.

"Feed your food." Those are the words to remember when you head to the kitchen counter or seedling station. At this point, the food your plants need is light, water, nutrients, and air.

Start with water. Check the moisture level of the soil your microgreens and seedlings are growing in. If the soil is not moist to the touch, add ½ inch of water to the bottom tray under your seedlings.

If any of your seedlings look leggy or weak, there's a good chance they lack sufficient light. Indoor plants need full-spectrum LED grow lights that are sited just a few inches above their tallest leaves. Be sure the plants get 12 to 16 hours of light each day.

Next, consider nutrients. If microgreens look yellow or lime green, there's likely not enough nutrients in the soil.

It's a funny thing to think, but plants also need air to grow. Use a small fan set on low near the seedlings, or be sure the seedlings are in a place with good ventilation. Airflow helps prevent fungal diseases and encourages stems to grow stronger.

Plants also need warmth to grow. Most cool season plants should be fine to germinate and grow in typical home temperatures of 60° to 70°F.

Use a humidity dome or a plastic cover to encourage seeds to sprout quickly, but be sure to remove the cover once the seeds sprout. Too much humidity can cause mold and fungal issues.

Day 5

This is the perfect day for troubleshooting any challenges with microgreens, sprouts, and seedlings.

If seeds aren't looking healthy, it might be the source that's the problem. Always use high-quality, untreated seeds specifically meant for sprouting or growing microgreens. Contaminated or poor-quality seeds can harbor harmful bacteria like salmonella or *E. coli*.

Next, keeping your growing area clean is important. Regularly clean and sanitize all equipment, including trays, containers, and tools, to prevent the spread of disease. A mild bleach solution or a food-safe sanitizer can do the job.

Microgreens and sprouts thrive in moist environments, but excessive moisture can lead to fungal diseases like damping-off (this looks like a thin, white mold growth). To prevent this, be sure there's plenty of air circulation around your plants and avoid overwatering. If possible, water from below to keep the foliage dry, and make sure your growing trays have adequate drainage.

Protect your plants from pests such as fungus gnats, aphids, or mites that can sometimes infest microgreens or sprouts. Fungus gnats, for instance, are attracted to moist soil and can be controlled by reducing watering and improving soil drainage.

Yellow sticky traps can help catch flying pests. For aphids or mites, a gentle rinse of the plant with water can help control these pests without harming the delicate plants.

Microgreens and sprouts require a balanced environment to grow healthily. Too much warmth can encourage mold growth and pests, while insufficient light can lead to weak, leggy plants. Maintain a consistent temperature suitable for the specific plants you are growing and provide adequate light, either natural or through grow lights, to promote strong and healthy growth.

Growing a variety of microgreens can not only be visually appealing and nutritionally beneficial but also can help in reducing the risk of disease spread specific to one plant type.

Give Me Five

Care for your seedlings in just a few minutes: Take out the seedling trays and check their water level. Test the moisture level with a soil moisture meter. Add a little extra water to the nondraining tray, if needed. Sprinkle earthworm castings on top of the soil in each seedling tray and double-check the position of the grow light above. You're done!

Day 6

Today's task is harvesting the best of what you've grown over the last week and so far this season.

Start with the hardiest herbs and greens and finish with microgreens and sprouts so you can make the most delicious meal and also save some of your bounty for the week ahead.

If you have woody herbs like rosemary, thyme, sage, and oregano growing indoors, take a few cuts from the plants. These herbs can be enjoyed right away or tied and hung in the kitchen to enjoy later in the week.

Herbs like parsley, cilantro, and chives growing indoors can be cut now to enjoy right away or stored in a glass of water in the fridge or on the kitchen counter.

If greens are growing under cover outdoors, you can take early cuts of kale, cabbage, or Swiss chard by harvesting the outer leaves first, allowing the younger inner leaves to continue growing.

Microgreens are ready to harvest just after the first leaves have appeared. This is typically 1 to 2 weeks after planting, depending on the type of seed. Lettuce and brassica seeds will be ready first, while parsley, carrot, spinach, and Swiss chard leaves will be ready last. To harvest, use a sharp pair of scissors or a knife and cut the microgreens just above the soil level. Be sure to wash them gently to remove any soil or seed husk. These tiny greens are perfect for a winter salad, where their intense flavors really shine. Toss a mixture of microgreens like arugula, radish, and beet greens with a light vinaigrette. Add nuts for crunch and perhaps sliced fruit for sweetness.

Finally, it's time to harvest your sprouts, which is as simple as giving them one final rinse, spinning them dry, and then eating them immediately or storing them in the fridge.

Stir chopped herbs into cooked grains or sprinkle them over roasted vegetables for a fresh flavor. Early cuts of winter greens can be lightly sautéed with garlic and olive oil for a simple yet delicious side dish. Sprouts can be tossed on top or used as a garnish on soups or stews.

While most of your neighbors are eating shipped produce from the grocery, you're making fresh, delicious dishes with food you grew yourself in just 5 minutes a day, even on the coldest days of the year. A warm quinoa salad with sautéed early cuts of greens, mixed in with a handful of microgreens and topped with sprouts, provides a nutritious and hearty meal. Another option could be a winter soup, garnished with fresh herbs and a sprinkle of sprouts for a burst of freshness.

Even though microgreens, sprouts, a few winter greens, and fresh herbs may just seem like a bunch of leaves, these green sprigs convert simple dishes into a gourmet treat. With a quick harvest, you've added loads of nutrients and taste to any winter dish.

Day 7

BJ Fogg, author of *Tiny Habits*, says a habit won't stick if you don't celebrate it. That celebration can be as simple as saying, "I'm awesome," clapping for yourself, and just smiling. Or you can take a seat, grab a fork, and eat your gardening reward one tasty bite at a time.

This day of enjoying the fruit of your labor is just as important as the labor itself. On this day, you rest and enjoy the work you've done by bringing in something fun just for you.

Perhaps you prepare some garden-to-table recipes ahead of time, enjoy a larger salad, invite friends or family over for a feast of microgreens omelets, avocado and sprout toast, mint tea, and homemade potato soup topped with microgreens.

Or maybe you celebrate by spending 5 minutes in a crafty way. While during the cold season it can seem like there's nothing to enjoy except the snowflakes, there's actually so much waiting for you outside, and there's comfort that comes from bringing the outside in. You could create a cold season arrangement with gatherings of evergreens, grass plumes, seed heads, and dried plants from the garden. Use berries, nuts, grasses, or dried flowers for color. If nothing else, bare branches from birch or willow trees can make a cold room feel warm.

Once you've brought the sights of the outdoors inside, bring some of the smells as well. Make sachets with dried herbs and flowers. Use lavender, rosemary, dried flower petals. Mix these with dried citrus peels or cloves and place in a small fabric sachet to tuck into tight spaces like drawers or closets where the fragrance will linger.

Mix up potpourri by adding essential oils to dried leaves and flowers and put it in bowls or small containers throughout the house. Or prepare a simmer pot with herbs and wood cuttings that release the smell of the outdoors inside.

DAYS

There are just a few secrets to surviving the cold season. Get outside at least twice a day, no matter how cold. Surround yourself with green, whether it's with baby plants, microgreens, or just some sprouts on the countertop. Nourish your body with garden-inspired drinks and flavors (even if it's from last summer). And allot plenty of time to plan and rest. In just 5 minutes, you can do all five of these things and live like a gardener, even if it's freezing outside.

*"I've been gardening for nearly a decade
as an adult, but always helped my dad in his
little garden plot growing up. I take 15 minutes
in the morning for my quiet time. I spend this
time in the garden for quick daily weeding,
tending, and harvesting.*

*"I schedule a walking break every hour
to rest my eyes, as I work from home. I use
that time to get in my garden and quickly plant
some seeds or starts, tend, weed, water, or just
observe what is happening in my garden."*

— Hannah

*"I walk through and prune my garden
with my morning coffee."*

— Jenna

Morning

As your day begins, you'll likely want something warm to drink and a few minutes to watch the sun rise, because the daylight hours are short this season. Take some time this morning to connect to the cold, the dark, and the full experience of the season you're in.

7:00—Make tea: Use herbs you've grown and dried, such as mint, rosemary, or lemon balm. This comforting drink reminds you of all you've already accomplished in the garden and encourages you to continue.

7:01—Journal: Take some time to write down the expected high and low temps for the day and list today's top goal for the garden based on the month and day. If temperatures allow it, plan to plant something outdoors. If it's too cold, wet, or snowy, make this the day you start a new tray of microgreens or tend to indoor starts.

7:02—Rinse sprouts: Run the tap, slide your sprouts under the water, and ensure all the sprouts get thoroughly rinsed. Rinse out the draining tray, put the top on, and slide the tray back to its spot.

7:03—Turn on the grow lights for microgreens and seedlings: Check the water level as you do.

7:04—Harvest: Snip microgreens from your seedlings.

7:05—Mix up: Prepare a quick breakfast with sprouts or microgreens.

REAL FAST FOOD

This is the season for sprouts, and there's nothing that makes you want to keep soaking and rinsing on any cold day than a delicious breakfast made from the sprouts you grew yourself. Here are a few simple recipes you can make for a garden-to-table breakfast in the middle of winter to keep you growing and warm all season long.

- *Alfalfa Sprout Omelet:* Whisk eggs and cook them in a skillet, then top with alfalfa sprouts, diced tomatoes, shredded cheese, and salt and pepper to taste.

- **Alfalfa and Avocado Sandwich:** Top whole grain bread with sliced avocado, alfalfa sprouts, tomato slices, cucumber slices, and cream cheese or hummus, and season with salt and pepper.
- **Microgreens and Goat Cheese Omelet:** Fill a fluffy omelet with goat cheese and a handful of microgreens for a gourmet breakfast.
- **Microgreens Avocado Toast:** Top whole grain toast with mashed avocado, a squeeze of lemon, and a generous layer of radish or pea microgreens.

You can also turn any herb into a tea. Boil up some water, then follow these suggestions:

- **Mint Tea:** Dried mint leaves (peppermint or spearmint) with a slice of lemon or a bit of honey for sweetness
- **Chamomile Tea:** Dried chamomile flowers mixed with dried lavender for a calming blend
- **Lemon Balm Tea:** Dried lemon balm leaves combined with dried mint for a refreshing twist
- **Sage Tea:** Dried sage leaves with a dash of lemon juice to enhance flavor
- **Rosemary Tea:** Dried rosemary leaves with dried lemon peel for a citrusy note
- **Thyme Tea:** Dried thyme leaves with dried echinacea for an immune-boosting brew
- **Lavender Tea:** Dried lavender flowers combined with dried chamomile or mint
- **Lemon Verbena Tea:** Dried lemon verbena leaves with dried hibiscus for a tart, fruity flavor
- **Basil Tea:** Dried basil leaves mixed with dried rose petals
- **Fennel Tea:** Dried fennel seeds blended with dried mint for a soothing digestive tea

How to Make Your Own Lemon Balm Tea

Lemon balm tea is not only delicious but is also known for its calming properties, making it a perfect drink for relaxation.

Begin by picking a handful of fresh lemon balm leaves from your garden. Choose the most vibrant and healthy-looking leaves for the best flavor. Gently wash the leaves under cool, running water to remove any dirt or insects. Pat dry with a clean towel, or let them air-dry for a few moments. You can leave the leaves whole or, for a stronger tea, chop them to release more flavor.

Heat fresh water until it reaches a rolling boil. The general rule is to use about 1 cup of water for each serving of tea.

Place the lemon balm leaves in a teapot or directly in your cup, using about 1 tablespoon of fresh leaves per cup of water. Pour the hot water over the leaves and let them steep for 5 to 10 minutes. The longer the leaves steep, the stronger the flavor.

Strain out the leaves before drinking. Add honey, sugar, or a slice of lemon for extra flavor, if desired. Savor your freshly made lemon balm tea. It can be enjoyed hot or you can cool it down for a refreshing iced version.

Noon

Midday is the perfect time to habit stack—use your lunch break as a reminder to connect to the garden.

12:00—Hydrate: Take a moment to savor refreshing sips of herb- or fruit-infused water. This could be a simple concoction of garden herbs like mint or rosemary in a glass of water.

12:01—Step outside: Take a quick, refreshing break from any indoor activities and reconnect with nature, even if you're at work.

12:02—Observe the weather: Take note of the current temperature, the intensity (or lack) of the sun, and any signs of weather changes that could affect your garden plans.

12:03—Quick pruning session: If you're at home, dedicate the next 2 minutes to pruning or tending seedlings. Cut off dead leaves, remove yellowed leaves, or thin a few plants to be sure each seedling has the room it needs to flourish.

12:04—Take a picture: Wherever you are, take a picture to capture the reality of your surroundings and help you notice the uniqueness of today.

REAL FAST FOOD

If you haven't had lunch yet, this is a perfect time to enjoy a sprout or microgreen recipe, like one of the following:

- **Turkey and Alfalfa Sprout Wrap:** Into whole wheat tortillas, place sliced turkey breast, alfalfa sprouts, sliced cheese (e.g., Swiss or cheddar), mustard or mayo, and lettuce and sliced tomatoes.

- **Alfalfa Sprout and Hummus Dip:** Combine finely chopped alfalfa sprouts with your favorite hummus and some olive oil and lemon juice, then garnish with paprika or cayenne pepper.

- **Alfalfa Sprout and Beetroot Salad:** Cook and slice beetroots, then top with alfalfa sprouts, feta cheese, roasted pecans, and balsamic vinaigrette dressing.

- **Asian-Inspired Alfalfa Sprout Salad:** Mix alfalfa sprouts, shredded carrots, sliced cucumbers, chopped red bell peppers, and toasted sesame seeds, then dress with a mixture of soy sauce, sesame oil, and lime juice.

- **Alfalfa Sprout and Egg Salad:** Combine chopped hard-boiled eggs, alfalfa sprouts, Greek yogurt, Dijon mustard, chopped green onions, and salt and pepper to taste.

- **Microgreen Pesto Pasta:** Blend basil microgreens with pine nuts, garlic, Parmesan cheese, and olive oil to make a fresh pesto. Toss with your favorite pasta for a quick, flavorful meal.

- **Microgreen Salad with Citrus Vinaigrette:** Combine various microgreens like arugula and beet greens, add sliced oranges or grapefruit, and toss with a citrus-based vinaigrette.

- **Microgreen Topped Pizza:** After baking your pizza, top it with a generous amount of microgreens for a fresh, peppery finish.

*"It's always the first thing I wander
through when I get home from work.
It's my oasis."*

— Izz

Evening

Use your time at the kitchen sink to prompt your evening routine. As soon as you've finished the dinnertime cleanup, begin your 5-minute gardener habit.

7:00—Rinse sprouts: Run the tap, slide your sprouts under the water, and ensure all the sprouts get thoroughly rinsed. Rinse out the draining tray, put the top on, and slide the tray back to its spot.

7:01—Boil water for garden tea: Into a teapot or mug, toss some dried herbs—anything from mint to chamomile, depending on what you have available. Top with boiling water. Making tea can be a soothing ritual, especially on cold evenings.

7:02—Check on your seedlings and turn off the grow light: Observe their growth, check the soil moisture, and ensure the seedlings are doing well. Then turn off their grow light to signal the end of their "daylight" hours. This mimics natural light cycles, which is important for plant growth.

7:03—Plan in your journal: With your tea steeping, take a minute to plan one specific part of your garden, a dish you want to make with the harvest, or a new space to cultivate. Jot down just a few ideas for the upcoming season, such as plants you want to grow, layout changes, or soil amendments needed.

7:04—Conclude with tea: Now strain your tea, take a deep breath, then sip. Write one line of gratitude in your journal, capturing a moment from the day that's worth treasuring.

"It has always been the best stress reducer
for me after work to go in the garden
and water the plants or pick something
to eat for dinner. To be able to grow
something and nurture it is incredible.
I felt relief getting my hands dirty.

"I've been gardening since 1990
from a condo and mostly containers
to a 5-acre ranch."

— Amy

Grow a Year-Round Supply of Kale

If you start in the cold season, you can grow a year-round supply of kale for your daily smoothie, even through the coldest and hottest parts of the year.

Begin in the middle of cold season, when it's still freezing outside, and start the kale seeds indoors.

In the first month of the cool season, move the kale plants into the garden. The plants start to take off even with a little frostbite. By the middle of the cool season, it's time for the first harvest and, of course, the first smoothie.

From this point, you can harvest a little each day to make another green smoothie. To keep the plants growing, just pull one or two leaves from each plant daily.

Once the weather warms, add trellises and vining plants, such as cucumbers or pole beans, alongside the kale plants to shade them. To fend off pests, use nasturtiums and chives near the plants to balance the pest pressure. But the truth is, there's enough kale for both you and the caterpillars.

Through the cool and warm seasons, just keep harvesting kale and making more smoothies. At their peak production, you can pick pounds and pounds of kale to freeze for use in the next cold season. By this point, your kale plants will have become a huge forest and the green smoothies will just keep coming.

By the second season, harvest all the leaves you can salvage, make a few more smoothies, and put the final amount of kale in the freezer so you can make a smoothie a day throughout the cold season to last you until the process starts all over again.

REAL FAST FOOD

Even in the cold season, you can eat food you grew yourself—and you can make it quickly because you've got the ingredients right at your fingertips. Try these sprout-inspired dishes with the fresh sprouts you've been growing in less than 5 minutes a day.

- *Vegetarian Sushi Rolls with Alfalfa Sprouts:* Fill nori sheets with sushi rice, alfalfa sprouts, sliced cucumber, and sliced avocado; roll and cut into slices, and serve with soy sauce for dipping.

- *Alfalfa Sprout and Quinoa Salad:* Combine cooked quinoa, alfalfa sprouts, diced bell peppers, chopped cucumber, and feta cheese, and top with lemon vinaigrette dressing.

- *Chicken and Alfalfa Sprout Pita Pockets:* Stuff a pita with cooked and shredded chicken, alfalfa sprouts, and sliced cucumbers and tomatoes; drizzle in Greek yogurt or tzatziki sauce, and season with salt and pepper.

PREPARE IN THE COLD SEASON

I'm not sure why they call it seasonal depression when they could simply call it "cold and sad."

Have you felt it? I know I have. When temperatures drop and the mornings and afternoons are both dark, I find myself feeling down and dark too. It can be so hard to crawl out of bed, put on my shoes, and push through the day.

But that's where the garden comes in. It beckons you outside, forces you to think forward, and even warms you up.

Science shows that good feelings come from moving our bodies, planning for the future, and eating energizing foods. You'll do all three of these, and so much more, when you start gardening 5 minutes at a time in the cold season.

Five minutes can change your day, change your perspective, and definitely change your mood. I'm not saying your seasonal depression will be cured for good, but I'm not saying it's impossible. This is for sure: you'll feel much better by spending just a few minutes each day outside, planning the garden, or even growing a little bit of green on the kitchen counter.

The small things add up over time. These few minutes can give you just what you need to cheer up, keep going, or even be a little more happy, at least as happy as one can be when it's dark and cold outside.

QUICK PICKS

Pick and choose from the following list when you need a quick idea or direction to make the most of any free moment in the cold season.

MONTH 1
Planting

- Take an overhead garden photo.
- Make a cool season planting plan.
- Make a warm season planting plan.
- Make a hot season planting plan.
- Make a second season planting plan.
- Plan the seeds you'll purchase.
- Shop for seeds.
- Order seeds.
- Organize old seeds.
- Start sprouts.
- Order lights.
- Set up grow lights.
- Soak sprouts.
- Rinse sprouts.
- Make a sprout salad.
- Make plant tags.
- Start a tray of microgreens.
- Harvest a salad of microgreens.
- Start perennial herb seeds.
- Order a container for an indoor herb garden.
- Create an indoor herb garden.
- Winter sow leafy greens like kale and cabbage.
- Winter sow flowers like pansies and violas.

- Winter sow root crops like beets and carrots.
- Winter sow onions and leek seeds.
- Plant large, leafy green seeds indoors.
- Start a second round of microgreens seeds.
- Make a planting plan for one bed of the garden.
- Organize new seeds with what you already have on hand.
- Test seed health by soaking a few to see what percentage sprouts. If fewer than 80 percent sprout, use fresh seed.

MONTH 2
Tending

- Water a tray of microgreens.
- Rotate a tray of microgreens.
- Water perennial herb seedlings.
- Rotate perennial herb seedlings.
- Water microgreens.
- Add a fan to the microgreens growing area.
- Add earthworm castings to microgreens trays.
- Plant new varieties of microgreens.
- Prune yellowing leaves away from microgreens trays.
- Give sprout containers a deep cleaning.
- Buy frost cloth.
- Cover the garden during extreme weather days.
- Make plant tags.
- Prune away old leaves on indoor garden.
- Use cloches on greens that need protection in cold weather.
- Deal with pests on indoor seedlings.
- Clear debris from soil in the outdoor garden.

- Prep the outdoor garden soil with a layer of compost.
- Monitor growth of winter-sown seeds.
- Add additional frost protection on the coldest nights.
- Prune away the lowest leaves on indoor seedlings.
- Check indoor seedling light levels and intensity.
- Check soil temperature with soil thermometer for indoor seedlings.
- Propagate cuttings from outdoor herb plants.
- Use an Oya or olla (clay watering pot) to help with watering your indoor herb garden.
- Take note of the growth of winter-sown seeds.
- Give a deep watering to outdoor seeds on the warmest day.
- Plan recipes to use microgreens and sprouts in the following week.

MONTH 3
Harvesting

- Harvest sprouts.
- Make a sprout salad.
- Make a sprout sandwich.
- Harvest microgreens.
- Make a microgreens omelet.
- Make a microgreens sandwich.
- Make a microgreens salad.
- Make garden tea.
- Tie up herb bundles.
- Create a fresh herb sauce.
- Make dried mint tea.
- Make lemon balm tea.
- Move dried herbs into a sealed container.

- Make a dry herb rub.
- Make herbes de Provence with preserved herbs.
- Create a tisane with lemon balm or anise hyssop.
- Top homemade soup with freshly cut microgreens.
- Make a sprout wrap for lunch.
- Host a winter solstice garden-to-table party.
- Make homemade focaccia with dried rosemary.
- Make rosemary tea.
- Make a kale frittata.
- Make a kale smoothie.
- Make a small kale salad.
- Roast beets and top with microgreens.
- Cut small harvests from winter-sown arugula.
- Cut small harvests from winter-sown spinach.
- Make a small leafy green salad.
- Make homemade sourdough bread and top with garden herbs.
- Make winter garden pesto with microgreens.

Cool Season

PLANT

---

*"The beginning,
as you will
observe, is in
your imagination."*

— NAPOLEON HILL

Pair of scissors in one hand and a bowl in the other, you step out the backdoor just as the sunlight starts to hit the edge of the driveway.

There are a million things to do inside, but this has to be done right this minute or it just doesn't get done.

You stop at the edge of the raised bed and kneel down, reach out across the waves of green leaves, and start cutting. One, two, three cuts of kale and you already have a bowl-ful. You reach past the kale and cut two huge leaves of Swiss chard, toss those in the bowl, and quickly snip three handfuls of spring mix lettuce.

The bowl is spilling over at this point, but you're not done yet.

You stand up and walk over to the next garden's edge. You bend there to snip some green onions and chives and stretch a little to the right to cut off a fistful of Italian parsley too. Just for fun, you snap off a few purple violas and toss them in the bowl. You stand up, brush off your knees, grab the bowl, and walk back inside.

Thanks to yesterday's rain, just a few splashes of water at the kitchen sink and the greens are clean. You lay out a tea towel to soak up the extra water droplets, pat each tender leaf dry, and then pull out the cutting board.

You chop the kale and Swiss chard, first the thick stems into chunks, then the leaves into thin ribbons. The lettuce is next and only needs a few quick motions of the knife. Then it's the green onions, chives, and parsley all in one big bunch chopped into the tiniest bits.

You quickly rinse the bowl from the garden, wipe it down with that same tea towel, and start stacking—kale, Swiss chard, lettuce, herbs, and, just for fun, the flowers.

Salad tongs help you toss this fragrant and colorful mix with vinegar, oil, and some local honey right before you load up the container for your family's lunch boxes.

You cover the bowl and place the rest of the salad in the fridge. "Mine!" you say to yourself, pretending the rest of your family might actually fight you for it.

A sunbeam comes through the kitchen window and hits you right in the eyes; the day has officially begun. But you're more than just awake. You're alive. It's officially the cool season.

Technically speaking, the cool season intertwines with both the cold and the warm seasons. There will be days that feel more like winter, and others that seem like summer. But the average will land somewhere right around perfect.

Temperatures can dip below freezing in the middle of the night and early morning. You'll see at least a little frost on plants and leaves when you wake up to hear the birds singing and making their nests. And then things heat up quickly during the day, sometimes by as much as 20° or 30°F, going from 35° to 65°F before dropping again in the early evening.

This is the season to plant and plant, and then plant some more. But there is also lots of tending and more than enough to harvest.

Begin the cool season with three goals you'll aim toward each and every day of this season. Set a goal for yourself to eat from the garden, to accomplish a certain feat in the garden, and of course, to enjoy it. The garden, after all, isn't just for the birds.

Daily cool season goals that only take 5 minutes:

- Walking in the garden each morning
- Eating a fresh garden salad every day
- Having hot tea each evening from the garden

*"I try to grab at least one thing from the
garden every day. Herbs, greens, and flowers."*

— Esther

*"After dinner, I hand water and enjoy
10 to 15 minutes of solitude in my garden."*

— Kelly

Cool Season at a Glance

MONTH 1: Planting the garden, starting warm season plants indoors

Week 1: Clear the garden.

Week 2: Prep the soil.

Week 3: Plant the garden border.

Week 4: Plant the interior of the garden.

MONTH 2: Tending cool season plants

Week 1: Feed plants.

Week 2: Support plants.

Week 3: Prune plants.

Week 4: Defend plants.

MONTH 3: Harvesting cool season plants

Week 1: Harvest herbs.

Week 2: Harvest small greens.

Week 3: Harvest large greens.

Week 4: Harvest roots.

Perennial, Annual, or Biennial?

Perennial plants come back from the root ball every year in the garden, as long as the cold season was not so extreme that it killed the root. Perennial plants are a staple in the garden because you buy them once and can divide and enjoy them year after year after year. Perennials are the first to come back and start growing in the cool season and are often the last to die off in the second cool season too.

Biennial plants grow in the garden for at least two full years before going to seed in order to reproduce. Biennial plants in the kitchen garden include Swiss chard, kale, parsley, mustards, carrots, and collards.

Annual plants grow from seed and then produce new seed within one season, generally within 90 to 150 days. Most kitchen garden plants are annuals. Annual cool season plants include arugula, radishes, peas, lettuces, cabbage, and beets.

WHAT'S GROWING OUTSIDE IN THE COOL SEASON GARDEN

In the first month of the cool season, your garden will be filled with new sprouts of perennial plants. Perennial plants generally go dormant in the cold season but are the first to appear when the soil temperature warms and the days get longer.

Herbs from the mint family like oregano, anise hyssop, rosemary, thyme, lemon balm, and sage also start showing first signs of growth, with oregano and lemon balm appearing first.

Edible landscape perennials like asparagus and rhubarb, and berry plants like raspberry, blackberry, and blueberry, begin to show the first growth of the new stems during the first month of the cool season. And, of course, chives come back year after year as perennials.

Biennial plants begin to grow faster around the middle of the first month, including kale, parsley, Swiss chard, and celery.

The annual plants that produce the first greens include plants in the onion family like garlic, onions, shallots, leeks, and green onions.

In the second month, the vegetable plants you planted start to fill the garden. This includes all the large plants, like kale, cabbages, mustards, collards, and Swiss chard.

Small plants such as lettuces, beets, radishes, and carrots will all be growing from seed at this point too. They'll be simply sprouts in the second month, but don't let those tiny leaves fool you.

By the third month of the cool season, your garden will be absolutely packed. Asparagus and rhubarb, tulips and hyacinths will appear around the edges of your kitchen garden. Planting beds will be packed along the perimeter with small perennial herbs as well as flowers like pansies, calendula, and chamomile; the inside of your beds will overflow with huge leaves of collards, kale, Swiss chard, and mustard as well as masses of small leaves, such as arugula, lettuce, radishes, and baby carrots.

While most of your neighbors will just be waking up to the fact that spring has arrived, you will be bringing in basket after basket of delicious and nutritious food that you grew yourself almost as soon as the birds started their spring song.

What to Plant Outdoors in the Cool Season

Leaves
Arugula
Broccoli
Buttercrunch
Cabbage
Cauliflower
Celery
Chives
Iceberg
Kale
Lavender
Lettuce
Mustard
Oregano
Romaine
Rosemary
Sage
Spinach
Spring mix
Swiss chard
Thyme

Roots
Beets
Carrots
Garlic
Leeks
Onion
Potatoes
Radishes
Shallots

Fruit
Fava beans
Snow peas
Sugar snap peas

MONTHS

Most cool season plants need 60 to 90 days to go from seed to harvest. So in the first month, you'll focus on planting, in the second you'll focus on tending, and in the third you'll focus on harvesting.

"Our garden is a daily 'extra room' and therefore an integral part of our life. We live in the very south of Germany where the sun is always warm, even in December or January. We can have breakfast or dinner outside sitting in the snowy garden if just the sun shines.

"But literally I keep it with a German saying, 'Nur die Harten kommen in den Garten,' which translates to 'survival of the fittest.'"

— Sarah

COOL SEASON/MONTH 1: PLANTING

No matter the season, the first month is always for planting. This is the time to dig in and add seeds and plants to your space so that you can start enjoying the moments among your plants as soon as possible.

Cool Season/Month 1/Week 1

The soil is warming up, the sun is coming out, and you can actually venture to the garden without a coat, a hat, and mittens. It's tempting to rush out with a packet of seeds and start planting already. But if you can resist, I promise it works much better to take a few steps to prepare your garden beds before dropping a single seed.

As someone who runs as fast as I can from chores, I hate to say these next two words to you, but . . . *clean up*. At the end of the cold season, a fair amount of debris has settled into your planting beds or lingered from your previous growing season. This needs to be cleared before you start planting, especially in areas where you'll sow seeds directly into the soil.

It's out with the old, in with the new. Spring cleaning. Tidying up. Use a simple rake to comb over the surface of the soil, leaving nothing from the previous season that could interfere with the new season's growth.

As you clear out old leaves and twigs, carefully remove any pieces of fruit you might find. I've left the shriveled skin of a dried-out 'Sungold' tomato in my raised bed and ended up with volunteer tomato plants popping up all over. You might think, "Cool! Free plants!" But many of these tomatoes are mystery hybrids that won't grow true to the original parent plant. It's much more work to pull them out later once they're established, so just do it now at the start of the season.

As you clean, keep an eye out for any perennials or cold season plants overwintering in your garden beds and avoid disturbing them. Be sure the location of the garlic cloves, leeks, and other greens you planted in fall are still well marked so you don't accidentally pull the good along with the spoiled.

Once the beds are cleared, top-dress them with a few inches of leaf mold or mushroom compost to restore the nutrients used up in the previous growing season and to serve as a clean slate for planting. Even if you make your own compost, you'll probably need to purchase some bags to completely cover your planting beds. I am diligent about tossing my lawn debris and kitchen scraps in my composter, but I still can't make enough homemade compost to meet the demands of my garden. That's why I turn to mushroom compost. One bag is sufficient to cover a 15-square-foot bed with a layer of compost about 1 inch deep. If you're gardening in a raised bed, add compost until the soil surface is level with the top of the raised bed.

As soon as you add the compost, you'll see the difference in the color and texture of your soil. Your planting beds will be fresh and ready to go! And you will have completed the "worst" spring planting job first. Thankfully, slowing down and doing the heavy lifting now means a lighter load for the rest of the season.

This is the first week of the planting month, and though warm weather may feel years away, this is also the time to purchase seeds for the warm season garden. As you plant loads of cool season plants outdoors, replace the empty spots on your indoor growing shelves with trays of seedlings for the warm season.

Listed here are the seeds you need to sow as soon as possible (if you haven't already) so the seedlings are ready for the warm season. Order seeds or bulbs/tubers or starts now so you can start them indoors before the end of the cool season planting month. This gives plants 90 days or more to grow to size for moving into the warm season garden.

"I garden in the morning. Coffee, then go outside, see the morning sun, listen to the birds, water and prune the garden."

— Brit

What to Start Indoors in the Cool Season (for the Upcoming Warm Season)

Leaves
Basil
Kale (if you miss
cool season)
Swiss chard (if you
miss cool season)

Fruit
Beans
Cucumber
Eggplant
Large gourds
Large melons
Okra
Peppers
Squash
Tomatillos
Tomatoes
Zucchini

Roots
Ginger
Sweet potatoes
Turmeric

Flowers
Calendula (if you
miss cool season)
Coreopsis
Cosmos
Marigold
Nasturtium
Petunia
Strawflower
Zinnia

Cool Season/Month 1/Week 2

Once the garden is cleared of debris and compost is added, check the soil's workability, temperature, and nutrients.

Workable soil is loose and porous—and ready to be planted. To test if the soil is workable, take a hori hori knife or a small trowel and dig down into the planting bed the full length of your garden tool. If you don't hit frozen ground, your soil is workable and ready to support the roots of your favorite herbs, vegetables, and flowers.

Now, using a soil temperature gauge, stick the tip 4 to 6 inches down into your soil to get a soil temperature reading. Cool season seeds, though able to tolerate frost, have a minimum temperature requirement for seed germination. Garlic, onions, and chives can start growing when the soil temperature rises just above freezing. Kale, mustards, broccoli, and cauliflower germinate when soil temperatures are around 45°F. Radishes, carrots, and beets can germinate in temps as low as 40°F. Lettuces, arugula, and greens start growing when the soil reaches 50° to 60°F.

Once you know your soil workability and temperature, test your soil too. A simple soil test can be performed by gathering samples of your garden soil and bringing them to a nearby agricultural center. Soil sample tests determine the nutrient makeup of your soil and give you advice for adjusting soil nutrient levels before you start planting for the year.

As you prep your garden beds, also prep your plants. This week, you begin slowly moving your plants outdoors for short periods of time. This process, called *hardening off*, allows plants to gradually adjust to the new temps and conditions outdoors so they don't go into shock when the big move happens. To harden off plants, move plant trays out to a partially sunny location once the coldest part of the morning has passed. Keep the plants in partial sun throughout the warmest part of the day and move them back indoors before the sun sets.

"The initial setup for the garden takes some time, but once that's done, you garden in small pockets of time in the morning and evening."

— Corina

If seeds for your warm season garden have arrived, this is the week to start those seeds indoors. You'll need draining and nondraining trays, soil mix, seeds, grow lights, and heating mats. Moisten the seed-starting mix so it is wet to the touch but not dripping wet. Place the soil blend in draining trays and plant one seed per cell. Barely cover the seed with a little bit of seed-starting mix, cover the tray, and place it inside a nondraining tray on a heat mat set at 65°F. Once seeds sprout, remove the cover and add water to the nondraining tray to ensure the new seedlings get all the moisture they need to grow to the next level. Keep seeds under the lights for 12 to 14 hours a day.

By the time you finish this week, you've tested and prepared the soil, begun to harden off plants to move out into the garden, and started seeds for the warm season indoors—all in just 5-minute increments.

Give Me Five

Got 5 minutes? You can do the following things this week:

- Test the garden soil's workability.
- Check the soil temperatures.
- Check the soil's moisture level.
- Take a soil test.
- Start to harden off your plants.
- Add more plants in empty spots.

Cool Season/Month 1/Week 3

The beds are prepped and cleaned, the soil is rich and warming up every day. It's finally time to plant!

I recommend planting out your cool season garden the Gardenary way: Borders first, then large plants, medium and small plants, and finally seeds. You plant from the outside in, filling in the blanks as you go. This approach makes the most of your space, increases production, and results in the most beautiful garden possible.

I love the ebb and flow of the garden, where some plants finish up while others take their first baby steps or reach their prime. But I also love consistent beauty in my beds during every season. And the way I make that happen is by planting the entire bed with both herbs and flowers—and I mean *a lot* of herbs and flowers.

My formula for determining the number of herbs and flowers in a garden bed goes like this: The perimeter of all the beds combined equals the number of herbs I'll plant. And the perimeter minus four equals the number of flowers I'll need. If you had two 4-by-8-foot beds, for example, you would need 48 herbs [2(4+8+4+8)] and 44 flowers [2(4+8+4+8) – 4].

Herbs for the Gardenary border usually include oregano, chives, thyme, rosemary, sage, marjoram, winter savory, and summer savory as well as chives and any other woody herb you'd like to grow.

For cool season flowers, plant snapdragons, pansies, violas, chamomile, and calendula between the herbs. Flowers not only make people happy, they add beauty to your beds, attract beneficial pollinators, distract pests that could otherwise harm plants, and help bring balance to the garden's overall ecosystem.

So before you plant anything else, plant the **border** of your kitchen garden. You'll find the Gardenary **border** offers consistent beauty and production to your **garden for** the entirety of the year while it also provides a wall of **pro-** tection for your crops from pests.

It's the first step to your garden's masterpiece.

Give yourself 5 minutes a day to work at planting **the** border. Take the first day or two to lay out all your **plants,** starting with the border in the same way you'd begin **with** the edge pieces of a puzzle or to frame a picture. Then **dig** in and start planting. Depending on the size of your **gar-** den, you may have the entire border planted in one **short** 5-minute session, or you may need several days. And **you'll** be amazed at the difference a fully planted border can **make.**

Give Me Five

Here's how to make the most of your 5-minute mom**ents** this week:

- Lay out the border of the garden.

- Plant herbs along the border.

- Plant additional herbs on the border.

- Plant one set of flowers along the border.

- Plant a second set of flowers along the border.

- Fill the empty spots and add more plants where needed.

- Take a picture and add any more plants neces**sary** for the border.

Cool Season/Month 1/Week 4

Your beds are bordered with herbs and flowers. Now, it's time to fill in the rest, 5 minutes at a time.

Biggest to smallest, that's the order for planting garden beds. Begin with the plants that grow the largest, then install medium-size plants, then small plants. Finish by sowing seeds.

We start with the largest plants to ensure those plants get enough resources. Give each one approximately 1 square foot of space to grow to its maturity. In the cool season, these plants include cabbage, kale, mustards, broccoli, Brussels sprouts, cauliflower, and Romanesco—as well as fruiting crops like sugar snap peas, snow peas, and fava beans. Peas need less horizontal space if you plant them along a trellis to grow tall instead of wide.

Next, fill in with any medium plants you have ready to go as well as including celery, Swiss chard, parsley, cilantro, or dill. These can be planted at least 4 per square foot.

Then, set in small plants like spinach, romaine, or other small lettuces at 6 to 9 plants per square foot—or even tighter if you expect to harvest from them frequently.

Finally, it's time to add small seeds in the remaining spots. Carrots, radishes, beets, arugula, lettuces, spinach, and any other leafy green seeds can be spaced from 9 per square foot (for beets) or 16 per square foot (for carrots).

By the end of this week, you've fully planted your cool season garden.

Start with Plant or Start with Seed Checklist

Ready to harvest in 60 days from planting?
Yes—seed   No—plant
Requires 1 square foot of space or more?
Yes—plant   No—seed
Germinates easily and quickly?
Yes—seed   No—plant
Requires more than 60 days from seed to harvest?
Yes—plant   No—seed

Give Me Five

Take 5 minutes a day this week to do the following tasks:

- Plant large plants.

- Finish planting large plants.

- Plant medium-size plants.

- Finish planting medium-size plants.

- Plant small plants.

- Plant seeds.

- Add more plants in empty spots.

COOL SEASON/MONTH 2: TENDING

April showers bring May flowers. If your cool season spans from March to May, you'll hope the old adage comes true for you. You'll need the rain because the seeds and plants waking up in their new homes are going to be thirsty.

This is the month to be certain your plants and seeds get the assistance they require to go from seedling to harvest. Trust that nature will be by your side, providing just the right amount of rain and sunlight.

But the one thing that never changes with nature is that it's always changing. Your job is to take up the slack when nature skips a bit, filling in the gaps and helping your plants thrive in the garden.

gardener time

"I've always loved gardening with my parents since a kid, but have been in a house with a garden to garden on my own for three years. I fit the garden into my day in my morning routine: Wake up, gratitude journal, and head to the shed to work out. Then play with my dog outside, check on my plants, and take some back inside to eat that day."

— Jen

Cool Season/Month 2/Week 1

Your cool season plants have just done something that plants don't like to do—move. Plants prefer to stay in place, and you've just forced them into a new home.

They'll need your help to make the adjustment. Water and nutrients are the key resources you need to provide now.

Water is never more critical than in the first 2 weeks after planting, when roots and leaves are in their most fragile state. Water the transplants right at the root level for the first 2 weeks. Don't let them dry out. Take a walk through your garden each day. Water when things look dry and hold off when plant leaves are soft and bright green.

Water is equally important for seeds. It penetrates the seed coat so that the plant embryo inside swells and eventually bursts through the coating. Until you see the first seedlings come through the soil surface, it's your job to ensure the soil doesn't dry out. If nature doesn't provide the rain, the watering job is on you. Each morning, use your finger or a moisture meter to test the moisture level of the top 1 to 2 inches of the soil. These seeds germinate near the soil surface, so you don't need to water deeply yet. Keep the area moist, but not soaking (too much water will rot seeds).

Even with all my years of gardening experience, I still can't tell you exactly how often or how much you'll need to water. This is because of evaporation rates and how often they change, not just in one season but especially in one area. Evaporation rates can change daily in the first cool season. Windy, sunny, or hot days call for more frequent watering. But cool and cloudy days mean you need to water less. So, take it from an experienced gardener: the person who knows best if it's time to water is the person standing there with the hose. So, take a minute every day to observe your plants' needs.

Beyond water, growing seedlings and plants may need an extra dose of nutrients. If leaf color shifts from dark green to a pale or lime green, or if seedlings grow to a point and then stop, it may be a sign that plants need a nitrogen boost. Spreading a bit of organic compost or a few sprinkles of earthworm castings around the base of each plant should provide what's needed to grow to the next level.

But if your plants show signs of slow growth, color shifts to light green or yellow, or your plants look stressed (and you know watering isn't the issue), consider a soil nutrient test (see page 86) to determine what supplements or treatments are needed.

Give Me Five

Make the most of your minutes—here's a short list for this week in the garden:

- Check moisture level.
- Thoroughly water seeds and seedlings daily in the first few weeks.
- Check color and health of seedlings.
- Add compost at the base of plants.
- Take an observational walk through the garden.

Cool Season/Month 2/Week 2

You wake up each day this month and see your plants are literally changing overnight—a little taller with a few more leaves. Now it's time to give them some support.

The simplest way to support new plants is by hilling, which is simply the practice of pushing a bit of soil or compost around the base of the plant. This ensures roots are not exposed, provides additional soil room for roots to expand, and creates a stronger base for the plant to grow from. Be careful not to hill the soil too high against the stem—just an inch of additional soil helps.

Support stakes are useful for large plants like kale, broccoli, and cauliflower. Install the supports this week so the plants can be tied to them as needed.

Tall trellises, panels, or arches with thin wire lines are ideal for vining plants like peas and fava beans. Add these structures now so you can help the vines attach as they grow.

Build an Obelisk Trellis

An obelisk trellis is a conical structure that holds vining and fruiting plants in the garden to maximize growing space and increase airflow. Creating your first obelisk trellis is as simple as gathering four to six tall bamboo poles and some twine. Here's the process:

1. Select poles that are at least 6 feet tall.

2. Sink the poles approximately 6 inches into the soil in the shape of a circle, ideally on a flat area that receives at least 8 hours of sun.

3. Using twine and working from the ground upward, create circles around the poles that the peas can hold on to as they climb.

Give Me Five

Here are some things you can make happen in less than 5 minutes this week:

- Check soil moisture level.
- Thoroughly water seeds and seedlings daily in the first few weeks.
- Check color and health of seedlings.
- Add compost at the base of plants.
- Take an observational walk through the garden.

Cool Season/Month 2/Week 3

After just one season in the garden, you'll realize that plants have a mind (and a motive) of their own. And what makes a garden a garden is your effort, a little each day, to direct and coax your plants to grow in line with your goals and motivators. In a nutshell, this means pruning.

Pruning entails cutting away a little to gain a lot. It directs a plant's energy. Each time you cut away one part of the plant, you send a message to the plant: less of this and more of that.

This month's first pruning step? Thinning seedlings. This is necessary to ensure each young plant has the space it needs to grow to its full potential. Cut back seedlings so that you end up with one radish for every 1 inch of soil, one carrot for every 2 inches, and one beet for every 3 inches.

When thinning, remember to keep the seedling that appears strongest with the thickest stem and broadest leaves. Trim the other seedlings at the soil level. This way, you give the healthiest specimen the best chance at growing to full capacity.

You shouldn't need to thin transplants, as long as you installed them with the correct amount of growing space. (See page 88 for a refresher on the Gardenary way.) But you may need to prune some leaves and stems.

First, remove any leaves that appear discolored or damaged from the move to the garden. It's perfectly natural for a plant's oldest leaves to slowly decline and stop growing. They've done their job to get the plant to a certain point of growth and are no longer necessary (because there's so many more leaves that have grown in their place). As old and damaged leaves are removed, the plant will focus on new growth.

Once you've removed damaged leaves, it's time to prune in order to direct growth. For large plants like cabbages or broccoli, this entails cutting leaves from the base of the plant to encourage the plant to grow tall instead of wide.

For peas, pinch off the tip of the plant when it's 6 to 12 inches high to encourage bushier vines. Alternatively, to spur longer vines, cut off side shoots from the main vine.

Give Me Five

Got 5 minutes? Here's your short list of garden tasks:

- Thin carrot seedlings.
- Thin radish seedlings.
- Thin beet seedlings.
- Prune old or damaged leaves.
- Prune outer leaves.
- Pinch back peas.

Cool Season/Month 2/Week 4

In the final week of tending, you'll defend and protect your seedlings and young plants from any insects or larger animals.

Using garden mesh is a simple, effective way to protect vulnerable plants—particularly young seedlings and delicate plants—from birds, insects, and larger animals. The mesh acts as a physical barrier, blocking pest access while allowing sunlight, air, and water to penetrate. Simply cover the garden beds or individual plants with the mesh material, securing it to stakes or frames elevated above the plants. The key is to install the mesh early in the season, before pests have a chance to establish themselves.

When it comes to pest protection, start by checking your plants regularly. Look under the leaves, near the stems, and around the base, where common pests like aphids, slugs, and caterpillars can often be found. Also check the surrounding soil area, especially for brassicas like kale, broccoli, and cauliflower. Early morning or evening is a good time for this inspection, as some pests are more active during cooler parts of the day.

Once you see pests, you can make a simple garlic spray that can help deter them without harming plants or beneficial insects. Crush a few cloves of garlic and mix them with about a quart of water. Let the mixture sit for a day or two, then strain the liquid into a spray bottle. You can apply this mixture directly onto the leaves and stems.

Another simple homemade control is an insecticidal spray made with castile soap. Simply mix the soap with water (typically at a 1:10 ratio) to dilute and spray on leaves affected by pests. (Note that I rarely do this, as I'd prefer to just remove the plant than spray.)

For best results, use both the garden mesh and a garlic or soap spray in tandem. The mesh provides immediate physical protection, which is especially useful for keeping larger pests at bay, while the garlic spray adds a layer of defense against smaller insects.

If pests continue to be a problem, consider interplanting more herbs and flowers to detract pests or to act as trap crops (plants that distract pests from food crops).

Give Me Five

When you've got 5 minutes, pick one of the following tasks:

- Check plants for pests.
- Make homemade garlic spray.
- Make homemade insecticidal soap spray.
- Spray affected leaves.
- Cover tender plants with mesh cloth.
- Check plants for pests after treating.
- Add trap crops like calendula and beneficial herbs like cilantro and parsley.

COOL SEASON/MONTH 3: HARVESTING

It's hard to believe this month has already arrived. It feels like yesterday you were longing for the first sunny days of the year and seeing the first sprouts of green.

And now, here you are, harvest basket in one hand and scissors in the other. It's time to cut and eat and cut some more (and eat some more too!).

All the minutes you've given to the garden this season have had a compound effect. The results are more than just additional—they're exponential. And you're about to see how.

Cool Season/Month 3/Week 1

This is the month you'll get to savor the taste of hard work.

The first harvests come from the herbs. Oregano, lemon balm, thyme, and chives are ready to be cut and used in cooking right away.

When harvesting oregano, look for bright green leaves and cut the stems a bit above where the leaves grow out to encourage more stems and leaves to grow in their place. Oregano has a strong, spicy taste that's great in lots of dishes.

Lemon balm's citrus aroma adds a fresh taste to spring dishes. Pick it in the morning when it smells the strongest. Use pruners to remove the top leaves or simply cut a few stems. Cut less than a third at a time to ensure the plant stays robust enough to grow more leaves.

With thyme, use scissors to cut the top part of the stem; that way, you'll come back to more leaves on new stems in just a few weeks. Thyme has a strong taste, and you only need a little bit to make a big impact on your favorite meals.

Chives are easy to pick. Cut the leaves about an inch from the ground and use right away. They taste a bit like onions and are great in salads.

"I am a telehealth therapist from home surrounded by gardens on all sides. Any break in my day is my opportunity to zoom out and work in my garden. My garden enables me to hold space for my clients so that I am refreshed and focused during our times together."

— Eve

Give Me Five

Got 5 minutes? Here are a few tasks you can check off:

- Harvest oregano.
- Cook a dish with fresh oregano.
- Hang oregano to dry.
- Harvest lemon balm.
- Enjoy lemon balm tea.
- Harvest thyme and chives and enjoy in a weekend meal.

Cool Season/Month 3/Week 2

There may be frost on the ground, but it's time to pick the first dark leafy greens. Kale, Swiss chard, mustard, bok choy, and cabbage are some of the first that are ready for harvest.

Kale is a green that can be harvested as soon as the leaves are big enough to eat. The trick is to pick the outer leaves first and leave the center ones to keep growing.

Small Swiss chard leaves are ready now. Cut the outer leaves when they're 6 to 8 inches tall from the base of the plant. New leaves will develop from the center of the plant in the weeks to come.

Mustard greens are quick growers. Pick the outer leaves when they are young and tender and let the inner ones grow. Mustard greens have a spicy kick, great for adding flavor to salads and stir-fries.

Bok choy, with its crunchy stems and tender leaves, can be harvested whole by cutting it at the base, or you can pluck a few outer leaves at a time. Bok choy is perfect for soups and Asian dishes.

Though cabbage is typically harvested as an entire head, you can begin harvesting the outer leaves of the plant right away for early spring salads and sautées.

Don't forget salad greens! At this time of year, garden-fresh salads will make you feel a totally new level of happiness, and watching these seed-grown plants multiply from one day to the next can only be described as magical.

Arugula is the quickest-growing green from seed, ready to harvest in just a few weeks. Pick the outer leaves when they are 3 to 4 inches long, and leave the rest of the plant to keep growing.

Spring mix, a variety of young salad greens, can be harvested when the leaves reach your desired size. Cut the outer leaves and the plants will continue growing from the heart.

Romaine lettuce takes a bit longer to be ready for harvest but is definitely worth the wait. As soon as the leaves form a loose head, you can pick the outer leaves or cut the whole head about an inch above the ground. If you leave the roots in place, you may get a second harvest from the same plant.

Endive and radicchio are often grown to a full head. Wait until the lettuces are firm and well formed, then cut the entire head off at the base. These greens add a nice bitter crunch to homemade salads.

Were the 5 minutes a day worth it? You'll know as soon as you take a bite of your first homegrown salad.

Give Me Five

Got 5 minutes? Here are a few tasks you can check off:

- Harvest kale and Swiss chard.
- Harvest mustard greens.
- Harvest bok choy.
- Harvest arugula and lettuce.
- Harvest spring mix.
- Harvest first spinach leaves.
- Make a delicious kale salad.

Kale Salad
(That You Actually Want to Eat)

I keep this recipe fairly simple (because that's how I do my kitchen time), but feel free to add or subtract based on what you can pull from your kitchen garden or find fresh at the farmers market. You might get fancy and top the salad with roasted Brussels sprouts, broccoli, or cauliflower and choose balsamic vinegar instead of cider vinegar. When you bite into this salad, you'll do so many good things for your body, and you'll be surprised by how much you suddenly love kale. I promise.

'Toscano' kale leaves
'Blue Curled Scotch' kale leaves
'Red Russian' kale leaves
Fresh radishes
Baby arugula or microgreens

Dried cranberries or cherries
Sunflower seeds or cashews
Local honey
Extra-virgin olive oil
Apple cider vinegar

Wash a few leaves of kale at a time. Discard any wilted or discolored pieces. (If leaves show signs of pest damage, I typically give them a good rinse and eat them anyway. Pest-damaged leaves are actually filled with nutrients.) Cut off and discard the stems and midribs of the three varieties of kale. Finely chop the leaves and place in a large bowl.

Using a mandoline, very thinly slice the radishes. Mix with the kale.

Chop the arugula and layer it on the kale-radish mixture. Top with the cranberries and sunflower seeds.

In a small bowl, whisk together equal parts of local honey, olive oil, and vinegar until well blended. Lightly dress the salad, mixing thoroughly.

Serve the salad right away, or cover and store in the fridge until the dressing softens the kale leaves and the flavors come together.

Cool Season/Month 3/Week 3

Leafy greens show off; it's easy to tell when they're ready for harvest. But root crops make you do a little digging. Since they're hidden, knowing when and how to harvest root crops is a bit of a guessing game.

Radishes are quick to mature, and are usually ready in just 3 to 4 weeks when their shoulders pop out just above soil level. Gently lift the radish leaves and sweep your hand along the soil line to feel for the radish. If the root looks and feels big enough to eat, gently tug on the leaves. Test one plant for size. When they're just right, pull them up and enjoy their peppery crunch.

Small carrots take longer to complete their growth cycle, usually 60 to 75 days. The tops of the carrots will start to show above the soil when they're ready. Gently pull on the tops to test their readiness. Harvesting is easier if you first use a garden fork to loosen the soil around the roots.

Beets are ready 70 to 90 days after planting; look for their shoulders above the soil line. As with carrots, loosen the soil around the roots to make pulling your treasure easier. Beets are great roasted or in salads.

Turnips and rutabagas take the longest of the root crops, usually 90 or more days. Wait until the roots are a couple of inches in diameter. Loosen the soil and pull them up. Turnips are milder and can be eaten raw or cooked, while rutabagas are sweeter and best cooked.

Give Me Five

Got 5 minutes? Here's what you can do.

- Test radishes for readiness.
- Pull radishes.
- Make a radish salad.
- Test beets and carrots for readiness.
- Pull baby carrots and beets.
- Make a beet and carrot salad.
- Test turnips and rutabagas for readiness.

*"Kick off the morning in the garden!
I pick Japanese beetles off my roses
while the beetles and I are still sleepy."*

— Min

Cool Season/Month 3/Week 4

So far this month you've filled your basket with leaves and roots. Now it's finally time for the fruit—the peas and fava beans growing on a trellis. Not only are these treats delicious, they're also so beautiful in the garden.

Harvesting peas at the right time is crucial for the **best flavor** and to encourage more growth. Snow peas are ready when the pods are 2 to 3 inches long and still flat; you'll be able to see the peas inside, but they aren't fully grown. Sugar snap peas are the sweetest when you give the seeds inside time to develop and plump, but pick the pods before the seeds become so large that they are hard and tough. To harvest peas, gently hold the vine with one hand and pull the pod off with the other. Bonus: the more often you harvest from the vine, the more peas the vine will produce.

Fava beans take a bit longer to mature than peas. The pods **are ready** when they are 6 to 8 inches long and the beans inside feel firm. Carefully pull or snip the pods off the vine. Picking fava beans regularly helps the plant focus its energy on producing new pods.

With a full basket, it's now time to enjoy the cool season bounty. Add herb-infused dishes, sauces, and spices to every meal. Easy roasted dinners of just-harvested radishes, beets, and carrots will be your favorite new side dishes. And there will be salads, lots and lots of salads.

Give Me Five

Got 5 minutes? Here are a few tasks you can accomplish:

- Keep harvesting kale and Swiss chard for a warm dish.
- Harvest lettuces and spinach for a salad.
- Harvest radishes to pickle.
- Harvest peas or fava beans.
- Stir-fry peas for a homemade pasta dinner.
- Harvest lettuces, carrots, and peas for a large salad.
- Harvest more peas to freeze.

Potato Planting Day

In the final month of the cool season, it's time to plant potatoes.

When you receive your order of seed potatoes, set the potatoes out in the light to "chit," or presprout. They are ready to plant once every spud has a little eye or two on its thick skin.

You may not be able to plant all your potatoes in just 5 minutes, but you can certainly plant a few each day, and the job will go faster if you're planting in a container.

First, prep your space: Fill a well-draining whiskey barrel or grow bag (a cloth bag used for growing plants), or amend a growing bed, with compost-rich soil. Place one potato, with the eye or sprout facing up, about 4 to 6 inches deep into the soil. Add another one, spacing potatoes about 6 to 12 inches apart. Be sure the potatoes are gently covered with 4 inches of loose, compost-rich soil.

Water the soil well, and as plants begin to grow, "hill up" by adding a little compost around the base of each plant to encourage the plant to produce more tubers and to protect the potatoes from sunscald. In a few months, you'll be digging up five to seven potatoes for every one you just buried.

WEEKS

As each day gets longer and lighter, your weeks will look different this season, but the structure of your 5-minute tasks can stay the same as it was in the cold season: plant early in the week, do some light tending in the middle of the week, and harvest at week's end. In other words, focus on plans and actions at the beginning of each week, take care of all you've done so far in the middle of the week, and finish the week by enjoying and harvesting whatever's grown so far.

"I make [gardening] part of my five-year-old's outside time. She collects pet grubs and I plant or weed!"

— Becky

Day 1

Temperatures and rainfall vary day to day in the cool season, meaning you should spend 5 minutes the first day of the week to check the weather for the coming 6 days so you can plan accordingly. Write down the predicted highs and lows and the expected days of rain. If temperatures rise more than 5°F within the next few days, plants may become stressed. If temps will drop below freezing for more than 12 hours, make plans to cover your newly installed plants.

Similarly, track how much rain (or snow!) your garden will get in the days ahead. If it's less than an inch of moisture per week, schedule a day to water. If lots of precipitation is on tap, you might need to protect your plants from overwatering or flooding.

As temperatures increase, your garden will need more water. Keep track of daily temperatures and adjust your watering schedule accordingly. Young plants and seedlings will need a gentle but steady supply of water to establish roots.

Spring can also bring windy days, which can dry out the soil, and you may need to water more often. Windbreaks or barriers can help protect delicate plants.

Certain plants thrive in the cool temperatures of early spring, while others need the warmth of late spring to grow. Use the forecasted temperatures to decide what to plant and when. For instance, leafy greens can handle cooler weather, but peas enjoy temps a little warmer.

Now that you know what weather is coming, you can make a plan for the remaining days of the week.

Day 2

Day 2 is perfect for planting. Use your 5-minute block of time to head to the garden with spinach or arugula seeds and add a new row alongside the one that's already growing. Or take a few minutes to seed a new tray of warm season plants like tomatoes or peppers.

Planting seeds at the start of the week is the perfect way to set intentions for the remaining days and gives you something to look forward to. You may return a few days later to find the first sprouts of arugula or spinach already showing up.

Arugula is one of the most carefree plants you can grow in your kitchen garden. To have a constant supply of arugula throughout the cool season, simply plant a new row of arugula seeds on day 2 *each week*. Use a ruler to set the seeds 2 to 3 inches apart in a long row. Water the area gently to moisten the soil without disturbing the seeds. If temperatures remain between 40° and 55°F, you should see sprouts before the week ends.

By staggering planting times, you create a cycle in your garden where new plants are beginning to grow just as older plants are ready for harvest. You'll have fresh arugula all season long.

Day 3

Successive planting of spring mix in your garden will ensure you never run out of fresh, tender greens.

First, plant spring mix seeds in a row, covering them lightly with soil. Water the area gently to moisten the soil without washing away the seeds. Wait 2 weeks, and then sow a new row of seeds adjacent to the first. As the season progresses, keep planting new rows of spring mix every 2 weeks. By staggering the sowing of new seeds every other week, you create a cycle of growth where new plants are sprouting and maturing just as earlier plants are reaching their prime, and you gain a constant supply of fresh greens, perfect for salads and garnishes.

Day 4

Today's the perfect day to head out to the garden with pruners, a harvest basket, and a compost bag. In just a few minutes, you can make the rounds and take a few snips to clean the garden, keep your plants growing, and even enjoy a few bites of fresh produce too.

Pruning is a simple practice that makes such a big difference in your garden's health and production. And take it from me, it's also a huge stress reliever. In the midst of a busy week, a few minutes in the garden with my snips and I suddenly feel like I can handle the world again.

Work on your large greens first, cutting away the lower and outer leaves. Keep the ones that look healthy and good to eat in your harvest basket and toss the ones that look a little spent in the compost bin.

Then head to the root crops. Not much pruning needed here; simply cut off any leaves that are discolored, misshapen, or filled with holes (that's a sign there's a caterpillar nearby).

Finally head to the peas, fava beans, and snow peas. If you're growing the vining varieties, pea pinching is in order—it's an easy way to increase your harvest. This method involves removing the growing tips of the plants to encourage bushier growth and then more pods. Pinching back, also known as *tipping* or *topping*, is a simple pruning technique. By cutting off the top few inches of the pea plant, you encourage the plant to put more energy into producing flowers and pods rather than just growing taller. This practice can lead to a fuller, bushier plant with more side shoots that develop more flowers and, eventually, more pea pods.

The best time to pinch back your vining pea plants is when they are 6 to 8 inches tall. At this stage, the plants are established enough to handle pruning but young enough that pinching back will have the most impact on their growth pattern.

To pinch back a pea plant, first identify the main stem and look for the growing tip at the top. This is where new leaves and tendrils are forming. Using a pair of clean, sharp scissors, cut off the top 1 to 2 inches of the stem, just above a leaf node (the point on the stem where leaves grow out). Be careful not to remove too much of the plant or damage the lower parts of the stem.

In a week's time, watch how your plants respond. In some cases, you might pinch back some of the new side shoots to encourage even more growth lower down on the plant.

Day 5

Devote a few minutes to feeding and protecting the garden today. Head out with two things: a bowl of earthworm castings and a bottle of garlic repellent spray (see page 98).

Take a quick walk through the garden and sprinkle a tablespoon or two of castings along each row of plants. If there's no rain in the forecast, follow up with a little drench of water along the roots of each plant (avoid the leaves, if possible).

With every row you pass, check to see if there's any sign of pests (like aphids, beetles, and caterpillars) or pest damage on any plant. If yes, spritz the plant with a little garlic spray.

Garlic spray is best used in the early morning or late afternoon to avoid the hot midday sun, which can cause the leaves to burn. Repeat this treatment every couple of weeks, or after heavy rain, to maintain an effective barrier against pests.

Now, in just a few minutes, you've fed your plants and protected them too.

Day 6

After a busy week of planting and tending, it's time to prepare a big cool season garden salad.

Gather your greens. For kale and cabbage, gently pull or cut off the outer leaves near the base of the plant. This way, the rest of the plant keeps growing. With arugula and spring mix, be more carefree: just snip off what you need, and the plants will sprout more leaves later.

Don't forget the herbs. Snip a handful of chives about an inch from the ground; they'll grow right back. For cilantro and dill, take the outer leaves and leave the rest. They're like the gift that keeps on giving.

Nasturtiums, pansies, or borage flowers are perfect for a spring salad. Pick a few for a pop of color and unique peppery flavor.

Before assembling the salad, wash all the greens and herbs to remove any dirt, but be gentle, especially with the flowers. Tear the greens into bite-size pieces and toss in a bowl. Chop the chives and sprinkle them in. Roughly chop and mix in the cilantro and dill for a burst of freshness and herby goodness. Now for the fun part: scatter the edible flowers on top.

Keep the dressing simple. Whisk together some olive oil, balsamic or apple cider vinegar (my favorites), a squeeze of lemon, a bit of honey for sweetness, and salt and pepper to taste. Drizzle it over your salad, give everything a gentle toss, and then have a seat and go ahead—give yourself more than 5 minutes to enjoy the best-tasting salad you've ever eaten.

Day 7

On day 7, prepare for the week ahead. And one of the best ways to do that is to harvest things you can keep fresh indoors for quick and easy use throughout the busy week.

Today's goal is a larger harvest of plants to keep you eating from the garden, even when your week gets busy.

Pick large heads like napa cabbage, romaine lettuce, or endive and store directly in the fridge to enjoy later in the week.

Pick sprigs of woody herbs like thyme, oregano, rosemary, or sage and place in a jar of water out of direct sunlight. These will keep for a few days until you're ready to use them.

Dig up root crops like beets, carrots, or radishes. Give them a quick rinse and wrap in a light dishcloth or paper towel; store in the fridge until you're ready to enjoy.

Harvest peas and fava beans, rinse these with a splash of cool water, and spin-dry or set on a towel. When fully dry, refrigerate in an airtight container in the produce bin.

You now have the foundation for a week of meals. Hold off on harvesting small greens as well as delicate herbs like cilantro, dill, or parsley until the moment you need them.

Beyond food, use day 7 to harvest a bouquet of cool season flowers to bring the garden to mind all week long, even when you're standing at the kitchen sink. Your bouquet may include hellebores, one of the earliest cool season flowers that come in shades of white, pink, purple, and green. For the best vase life, cut hellebores when the flowers are mature and fully open.

Forsythia branches, with their bright yellow flowers, are excellent for adding height and drama to arrangements. Cut forsythia stems when the buds are swollen but not yet open. Once indoors, the warmth will encourage the buds to bloom with a burst of yellow, a hint of the bright spring sunshine on its way.

Just as the weather starts to warm, daffodils appear. Cut a few bright yellow, white, or orange flower stalks that are slightly opened but not fully bloomed. Keep daffodils in a vase all by themselves, since their sap can affect other flowers.

If you want your home to smell like spring, cut a few hyacinth stems just when the lower flowers begin to open. Hyacinths work well in short vases and will make you feel like you're standing in the middle of your garden even when you're stirring at the stove.

Tulips bloom a few weeks to a month after daffodils. Leave lots in the garden for the bees and for the beauty, but bring a few indoors. Cut stems of flowers that are just starting to open. Place in a vase and put in a spot without direct sunlight so the blooms last as long as possible.

Magnolia branches can be cut when the buds are set but not yet open. These stunning blooms, often in whites and pinks, can be a focal point in any arrangement. Magnolias are best displayed in large vases and can bring a touch of elegance to your indoor decor.

Pansies and violas are so fun and one of the easiest flowers to grow in the cool season garden. These cheerful, colorful flowers can be floated in bowls of water for a whimsical and charming display. Both pansies and violas have a wide range of colors and patterns, allowing for creative and vibrant indoor decorations.

To get the most of your cool season flowers indoors, take cuts in the morning when the plants are most hydrated. Use sharp scissors or pruners to make clean cuts, and place the flowers in water immediately.

Changing the water every couple of days and trimming the stems can prolong their life and keep your arrangements looking fresh and beautiful.

Garden-to-Table Carrot Juice

Makes 4 servings

Carrot juice offers nearly all the benefits of garden-fresh carrots: beta-carotene and vitamin A (protect against age-related eye disease), vitamins C and B6 (immune boosters), and potassium (regulates blood pressure). The one thing you're not getting as much of when you drink carrots is fiber. You can use any color carrots in this recipe—white, yellow, purple, or the classic orange—but choose the sweetest ones. Make this juice in a blender or food processor. You'll also need a nut milk bag or a sieve.

About 4 fresh carrots, peeled (optional) and quartered, to make 1 to 2 cups

½-inch piece of peeled fresh ginger, or more as desired

1 lemon, quartered

Filtered water

Pinch of ground turmeric (optional)

Pinch of freshly ground black pepper (optional)

Place the carrots and ginger in a blender or food processor. Squeeze juice from the lemon quarters into the blender. Cover the mixture with filtered water. Add the turmeric and pepper.

Blend at low speed for 30 seconds. Slowly increase to high speed until the mixture is completely pulverized into a liquid.

Pour the liquid into a nut milk bag or sieve and strain into glasses. Enjoy this juice right after it's made.

> *"Gardening is part of my morning routine.
> Workout, then garden. This centers me."*
>
> — Stephany

Pea Planting Day

Pea planting day is one of my favorites. It's my form of sticking a flag in the sand and declaring that warmer weather is officially on its way.

Peas can be planted directly in the garden soil or started indoors. To start seedlings indoors, plant the peas a few inches down in deep seed-starting cells that can be easily opened. Place the cells in a nondraining tray with an inch of water. Set the tray under grow lights until the peas are 2 to 3 inches tall. At that point, harden off the seedlings by bringing the cells outdoors for increasingly longer amounts of time to expose the plants to sunlight and cool season air temperatures (see page 143 for details about hardening off).

To grow peas from seeds, soak the seeds in water overnight just before planting day.

On pea planting day, install a 6-foot-tall trellis. Along the edge of the trellis, dig holes 2 to 3 inches deep for seeds or 4 to 5 inches deep for seedlings. Plant each seed or seedling, fill the holes, and cover the area with 1 to 2 inches of compost. Gently water the planted area and keep the soil moist over the next few weeks as the seeds germinate or the young plants get established.

DAYS

In the cool season, you can view your *day* in three parts—but flip the order: the morning is for harvesting, the midday is for tending, and the evening is for planting.

In the morning, step through the garden to find things for dinner. Maybe it's just a few clips of parsley or basil to add as a garnish; perhaps it's a harvest basket full of tomatoes and peppers. Let the end of the day be a time to enjoy the results of all your hard work.

If you're home at midday, take a minute to stroll through the garden with pruners or scissors and just clean up a little. Trim a little here or there, or pull up plants that are no longer productive.

In the evening, consider bringing a few seeds out with you as you walk through the garden. Extra pea, arugula, spring mix, or cilantro seeds can be popped into the soil as you walk along the pathways and observe what's growing.

"I love to take my girls outside after breakfast and tend the garden while they play and swing."

— @hroberson529

Morning

As your day begins, it's the perfect time to drink some garden tea and enjoy the first light in the garden.

7:00—Start with a hot cup of tea: Use herbs you've grown and dried, such as mint, rosemary, or lemon balm. This comforting drink reminds you of all you've already accomplished in the garden and encourages you to continue.

7:01—Journal your top gardening goal: Cup of tea in hand, grab your journal and write down the expected high and low temps for the day and list today's top goal for the garden based on the month and day. If temperatures allow it, plan to plant something outdoors.

7:02—Give sprouts a quick rinse: Do this over the sink.

7:03—Turn on the grow lights for next season's seedlings: Check the water level as you do.

7:04—Head out to the garden: Take a quick walk and cut a few things for breakfast.

7:05—Make a quick breakfast: Yours might include kale, Swiss chard, green onions, and parsley.

Morning may be the most important segment of the gardener's day simply because nature makes it so. In the night, plants stop the photosynthesis process but continue respiration, that is, turning oxygen and glucose into carbon dioxide and water. Plants still grow overnight as they use the energy they gathered during daylight hours. They also store up sugars produced during the day.

This means that in the early morning, the leaves in the garden are the sweetest and crunchiest they'll be all day. Harvests gathered in the early morning are likely to be juicier and last longer than those made even a few hours later. No wonder morning is the best time to step outside with a pair of scissors or pruners and fill up your favorite bowl!

Lettuces prove the rule. These greens may wilt and take on a bitter taste in the high heat of the afternoon sun. But by early morning, these leaves have regained their water levels, crispiness, and sweet flavor.

Morning is also ideal for watering the garden. Put the hose on the shower setting and water the soil surface of each bed (not the plant foliage). Even if your garden has drip irrigation, a few weeks of watering the soil's surface in the morning is critical to ensure seeds get the moisture they need to germinate.

Cool Season Drinks

I began drinking a green smoothie every morning around the time I scored my first success growing dark, leafy greens. Once I started harvesting kale, Swiss chard, and spinach, I needed recipes to help me use up all that green goodness.

I've maintained this practice for over five years for a seriously healthy start to my day. I harvest fresh kale leaves early in the morning for as many months of the year as possible.

When I have surplus kale, I freeze the leaves, or I dehydrate and grind them into a powder to use in winter months. What's great about making green smoothies (besides the obvious health benefits, like all the fiber and omega-3s) is that it's a great way to use up leaves that have grown too large or are too tough for salad.

My Go-To Kale Smoothie

Makes 4 servings

I was inspired to make green smoothies by Jen Hansard of Simple Green Smoothies. Over time, I've tweaked my recipe to come up with my favorite version, although I still make changes based on what greens and fruit I have on hand. It's easy enough to make this recipe your own based on your personal preferences or whatever food color your toddler has decided is acceptable that day! Prepare this using a high-powered blender.

1 to 2 cups curly kale, dinosaur kale, or red kale (whatever type you have)

1 to 2 cups water

1 avocado, scooped into chunks

1 banana

1 cup frozen mango

Juice of ½ lemon

2 tablespoons pepitas (pumpkin seeds)

Give the kale a quick wash. There's no need to remove the center stem (that's where a lot of the fiber is!).

Place the kale in a high-powered blender and pour in enough water to cover the leaves. Blend on high speed *before* you add anything else. (This ensures the final result is nice and smooth—no large green chunks to get caught in your teeth.)

Add the avocado, banana, mango, lemon juice, and pepitas. Blend on high for a couple minutes. Enjoy immediately.

*"Now that I work from home a few days a week,
I'm able to pop out to the garden on my lunch break.
Sometimes I just do a quick survey or pull
a few weeds to get my hands in the dirt.
The best is when I can harvest my lunch or
dinner or just snack right then and there."*

— Jennifer

Noon

Use your lunch or midday break as the prompt for a 5-minute routine like the following:

12:00—Hydrate: Take a moment to hydrate with a refreshing sip of herb- or fruit-infused water. This could be a simple concoction of garden herbs like mint or rosemary in a glass of water.

12:01—Step outside: Take a quick walk outside or, if you're home, into the garden. This is your chance for a break from any indoor activities and to reconnect with nature.

12:02—Observe the weather: Take note of the current temperature, the intensity of the sun, and any signs of weather changes that could affect your garden.

12:03—Quick pruning session: If you're at home, dedicate the next 2 minutes to pruning a plant or two. Cut off dead leaves, remove yellowed leaves, or thin a few plants to be sure each seedling has the room it needs to flourish.

12:05—Make something: Bring in a few of your clippings to display in an arrangement or to cook or chop up for lunch or dinner.

Lunch is the best time to refuel yourself with a salad made with garden greens. If you've never had a freshly cut salad made with your own-grown greens, I might argue that you've never truly had a salad.

#Sixmonthsofsalad was the challenge I set myself at the start of 2020. Little did I know when I wrote my New Year's intentions what "new" things that year would bring. I planted seeds the minute that soil was workable in my Chicago garden, and by early April, I was cutting beautiful, sweet, and succulent leaves every single day.

When the pandemic shut down the world, and restaurants and stores closed, I was eating the most gourmet salads right from my yard. Though I was growing in less than 90 square feet, I knew I could make my #sixmonthsofsalad plan work because of the cut-and-come-again nature of salad plants: the plants I harvested from one day would be ready for another trim a few weeks later.

Marking your midday with greens from your own garden is the perfect way to establish a springtime or cool season routine. If you're not home during the noon hour, take time in the morning to cut the leaves you need for your lunchtime feast. Then rinse, dry, and pack your salad for the road. In the middle of a busy day, you'll get to open up that container, take a deep breath, and be transported right back to your earthy haven as you take one bite of salad after another.

Challenge yourself to a salad a day from the garden during this peak season. The greens are young and sweet, and no grocery store or farmers market can rival the taste you'll experience when you eat greens within minutes of harvest.

My #sixmonthsofsalad challenge taught me that eating freshly cut salad from the garden wasn't possible just in warm climates—it could work everywhere. To have your own 6 months of salad, just follow this simple formula:

1. Plant microgreens and sprouts weekly.

2. Plant large salad plants (including cabbage, kale, iceberg, romaine) every 90 days.

3. Plant medium-size plants (including Swiss chard, parsley, cilantro, dill, romaine) every 60 days.

4. Start seeds for small plants (arugula, spring mix, 'Black Seeded Simpson' lettuce, spinach) every 30 days.

Each week you'll harvest microgreens and sprouts to "top off" your fresh garden salad. Within months of planting large salad plants, you'll get to harvest the outer and lower leaves regularly for salad. Small plants start producing 45 days after planting, and you can harvest from them more frequently in the days that follow.

So if you happen to be home during the lunch hour, take a moment to step outside, stretch, and have a few minutes away from work, the screen, and sitting. Bring along a pair of pruners or scissors and walk through the garden to do a little tending.

Gardening midday reveals how your plants fare in the hottest and brightest part of the day. You'll notice if leaves are wilting, if the soil looks too dry (or too wet), and if parts of the garden need attention.

Tending in the middle of the day is less about doing hard work and more about observing—although if you're like me, you'll find that when you look, you can't help but touch. And as soon as you touch (maybe pluck a leaf, pull a weed), you are tending. Tell yourself you're just taking a walk, but come prepared with pruners!

Evening

Use your time at the kitchen sink to prompt your evening routine. As soon as you've finished the dinnertime cleanup, begin your 5-minute gardener habit.

7:00—Rinse sprouts: Run the tap, slide your sprouts under the water, and ensure all the sprouts get thoroughly rinsed. Rinse out the draining tray, put the top on, and slide the tray back to its spot.

7:01—Boil water for garden tea: Into a teapot or mug, toss some dried herbs—anything from mint to chamomile, depending on what you have available. Top with boiling water.

7:02—Check on seedlings and turn off the grow light: Observe their growth, check the soil moisture, and ensure the seedlings are doing well. Then turn off their grow light to signal the end of their "daylight" hours. This mimics natural light cycles, which is important for plant growth.

7:03—Plan in your journal: With your tea steeping, take a minute to write in your garden journal. Use this minute to note what's working or not working in the garden right now and what new plans come from that.

7:04—Step outside: Take one last stroll outdoors into the garden, brush off things that may be in your way, maybe make space for more seeds tomorrow, or take a few cuttings you can bring to arrange indoors.

7:05—Conclude with tea: Now strain your tea, take a deep breath, then sip. Write one line of gratitude in your journal, capturing a moment from the day that's worth treasuring.

"In more recent years, I fit in an evening stroll through the garden. My teaching job has become so stressful, and these walks are a release for me. I need this evening walk right when I get home and before I do anything else. It settles my overstimulated brain from my day in the classroom. This habit feeds my soul."

— Marie

The main task of the evening is to plant. Transplants may struggle if they are placed in garden soil during the heat of day. Planting them in the evening gives them time to settle in before they face the sun.

Stepping through the garden as the sun lowers also allows you a moment to replace plants pulled earlier in the day or week or to seed a few gaps. We're not talking heavy work; just wear an apron with a few packages of seeds and a spacing ruler or dibber in the pocket and pop in a few more peas or fava beans or perhaps a few green onions.

Evenings can also be the best time to start new seeds indoors or to nurture those seeds you've been growing indoors for a while.

And in the first month, you'll spend a few of your evenings moving plants back indoors during the hardening-off period.

Another beneficial evening task is deadheading, that is, snipping off old, spent blooms. Deadheading encourages plants to produce more flowers. It's a quiet, almost meditative activity that can be incredibly satisfying as the gardener, even as it helps prolong the flowering season.

Every sunset occurs a little later than the one before in the first cool season of the year, so make it a point to slow down and watch the sun drop below the horizon each night. Perhaps you set an alarm at the beginning of each week to remind you to watch; notice how you have to reset the alarm in the days ahead to echo the changing arc of the sun.

When you come back in the house, quickly jot down one observation about the garden. This may feel like journaling, but it's actually a combination of research, notes for next year, and gratitude all in one.

PLANTING IN THE COOL SEASON

It's been said that the anticipation of a major event creates the highest levels of dopamine in our minds. Dopamine is a type of neurotransmitter that our body makes to send messages between nerve cells. It plays a significant role in pleasure, motivation, and learning.

In other words, dopamine helps us feel good.

Most of us assume that the way to feel good is to get to the reward as fast as possible, whether that reward is the vacation, the job promotion, or, in our case, the garden harvest. But science shows that the most impressive dopamine release happens not at the moment of gratification or reward but during the anticipation phase instead.

Dopamine hits hardest not when you land in your vacation spot but during the days you planned, packed your bags, and charted your course; not when you finally get the job you always wanted but instead when you're working away at your computer, solving problems, and showing up at meetings along the way.

During the cool season, you may assume the goal is your reward. You're waiting for the soil to thaw, the cabbage leaves to tighten, the carrots to grow deep roots, and the pea plants to develop pods.

But each time you check on your seedlings or prune back a leaf (or two or three), your brain is quietly celebrating, slowly dropping a little more dopamine into your nervous system and telling you to keep going.

We might think that the garden is all about the harvest. But, really, it's the planting and tending that brings the biggest reward. The good news is that you don't have to wait all season to enjoy the garden—because the dopamine, the rush, the relaxation, the anticipation is here, waiting for you, each and every day of this season. To some people, gardening may look like a lot of waiting, but you know that you're in the midst of truly living—5 minutes at a time.

QUICK PICKS

Pick and choose from the following list when you need a quick idea or direction to make the most of any free moment in the cool season.

MONTH 1

Planting

- Order warm season seeds.
- Plant more microgreens.
- Soak seeds for sprouting.
- Rinse sprouted seeds.
- Plant onion seedlings.
- Order seed potatoes.
- Chit (presprout) seed potatoes.
- Cut up seed potatoes.
- Plant seed potatoes.
- Order seed ginger.
- Start seed ginger.
- Plant flower border.
- Plant herb border.
- Plant large plants.
- Plant medium-size plants.
- Plant small seeds.
- Plant peas.
- Plant calendula seeds.
- Start nasturtium seeds indoors.
- Grow basil indoors.
- Order spring garlic.
- Plant spring garlic.

- Plant strawberry plugs.
- Plant carrot seeds.
- Plant arugula seeds.
- Plant spinach seeds.
- Start warm season flower seeds indoors.

MONTH 2

Tending

- Measure soil temperature.
- Get a soil test.
- Test the pH of your garden soil.
- Build obelisk trellises.
- Make panel trellises.
- Build arch trellis.
- Get plant stakes and supports.
- Prune back native plants and shrubs.
- Trim asparagus spears.
- Trim rhubarb leaves.
- Pinch back pea plants.
- Thin root crops like radishes and carrots.
- Amend soil with earthworm castings.
- Create homemade pest spray.
- Clear debris from the garden.
- Start a compost pile with clearings from the garden.
- Install a rain gauge.
- Observe the rain gauge and empty it.
- Test your soil's moisture level.
- Mark the sunrise and sunset of the week ahead.
- Move plants outdoors for hardening off.
- Move plants indoors for hardening off.

- Install a rain barrel or way to save rain water.
- Trellis peas.
- Create garlic spray for pests on greens.
- Cut away lower and outer leaves from large greens.
- Add compost to the bottom of large plants.
- Hill soil around pea plants.
- Add phosphorus-rich fertilizer around pea plants.
- Tie up cabbages, cauliflower, and broccoli.
- Cut away pest-affected leaves.
- Spray homemade spray on pest-infected leaves.
- Prune herbs to make room for root crops.

MONTH 3
Harvesting

- Make a sprout salad.
- Make a microgreens salad.
- Take first cuttings of herbs.
- Make herb fresh dressing.
- Cut first greens from spring mix.
- Cut outer leaves of cabbages for Asian-inspired salad.
- Cut lower leaves of kale.
- Make a kale smoothie.
- Make green juice.
- Cut first batch of spinach.
- Make a spinach salad.
- Make spinach smoothie.
- Make a spinach dip for an appetizer.
- Make herb-infused olive oil.
- Make a dried herb salt.
- Make chimichurri with spring parsley and garlic.

- Create a green pasta sauce with spinach, garlic, and parsley.
- Complete a large harvest of cilantro, parsley, and dill for freezing.
- Start fermenting cabbage leaves for sauerkraut or kimchi.
- Harvest radishes.
- Roast radishes.
- Make radish pickles.
- Harvest beets.
- Roast beets with olive oil and sea salt.
- Make carrot and pea soup.
- Harvest peas.
- Cook peas in a stir-fry and serve over rice.
- Make garden-inspired tea and tisane.
- Save cilantro and dill seeds for spices.
- Host a spring harvest party and dinner.

"Herbs, herbs, herbs! Nothing takes me from busy to refreshed faster in my garden than picking a few herbs to add to a meal. A little basil on a lunch salad, rosemary for a roasted chicken, and fresh oregano on pasta sauce—it takes less than a minute and makes my mind, heart, and body so happy.

"When a friend or neighbor needs a boost, there's nothing like fresh flowers—except when I don't have time to go get them from the store. I keep old mason jars in my cupboard just to cut fresh flowers from my own garden. I feel refreshed in the act of gathering them and so grateful in the giving! Oftentimes, I'll make 'kitchen bouquets' with the inevitable abundance of basil, rosemary, and oregano. They are beautiful in a vase!

"I have a thing for bees. On a day I feel particularly overwhelmed with tasks, stepping out into the garden to follow a bee from flower to flower melts my stress and slows my pace."

— **Lara Casey Isaacson,**
founder of *Southern Weddings* magazine
and creator of PowerSheets goal planners

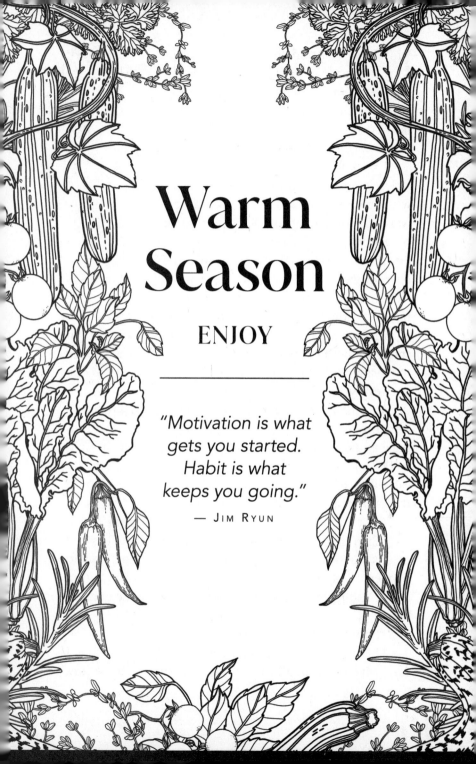

Warm Season

ENJOY

"Motivation is what gets you started. Habit is what keeps you going."

— JIM RYUN

I hit the stop button on my watch and slow my steps as I round the corner. I go from jogging to fast walking to strolling in a matter of a minute, wipe my brow, and take a deep breath.

Now comes the easy part.

I walk up the driveway and take a sharp left to head into the path overrun with lemon balm and find my way past the waist-high anise hyssop. My feet find the stepping stones, and it's only a few more steps until I'm standing under the tomato arch and leaning over the garden bed to break off at least four giant leaves of 'Toscano' kale. Then I pull a few of the green onions growing alongside the edge of the bed. The hot peppers are also within reach, so I pick just one of those. I can't help but smell the basil, so I grab a few leaves. Then I reach above my head, pick four or five cherry tomatoes, and try to balance everything in my arms as I head indoors.

I'm still breathing heavily from that morning run, and each inhalation brings wafts of pungent, spicy, sweet produce and soil. After sliding off my shoes, I head straight to the kitchen. It's time for a garden omelet, a meal I've earned with that long run and all the minutes I've spent in the garden up until that moment.

As butter melts in the skillet, I chop pepper and tomatoes, then cut up kale, green onions, and basil. The chopped greens are tossed into the pan first. In go the tomatoes and pepper, followed by two or three whisked eggs and a dash of salt and pepper. I might add some freshly grated cheese before gingerly folding the greens-heavy omelet in half and sliding it onto a huge dinner plate.

This garden-to-table moment is steaming hot, but I can't dive fast enough into this gooey, yummy goodness.

This meal has come full circle, from tiny seeds I planted months ago to stems and leaves and fruit I've tended for weeks. Now, this plate full of nutrition fuels me after a long run, making it possible to head back outside and keep planting.

The warm season is the ultimate compound effect. The minutes, hours, and days of sunshine, rain, compost, cleaning, pruning, and watering turn it into a sum far greater than its parts.

I call it the "Quintessential Garden Season" because it's the one we all imagine when we think of a garden—a harvest basket brimming with tomatoes, beans, peppers, squash, and watermelon. These fruits and vegetables thrive in the warm season, the period between the last frost of spring and the first one of autumn.

Did your grandmother ever tell you that you reap what you sow? Well, you'll find out what she meant in this season.

Every seed you tucked in the soil, every tiny seedling you nurtured, every leaf you pruned: your labor comes to fruition, literally, in this season.

And your most important job? To *enjoy* every single bite.

gardener
time

*"I hand water in the mornings before work.
It's so still and quiet; my favorite part of the day."*
— @growing.calm

*"After the kids get on the bus, I walk outside
and say good morning to the garden. I spend
5 to 10 minutes seeing how it did in the night.
I might run my fingers through the rosemary
and thyme and take a deep breath in. I'll take
a photo and capture the sun glistening off
the dew on the dill. I hear the birds sing.
I then go inside to start my day.
(Or was that the start???)*

*"When lunchtime hits, I venture back out
and say hello, for now my garden is in the full sun.
I drink in the warmth and maybe clip some lettuce
for lunch. I then come back inside and continue
my day. And after dinner is done and dishes are
cleaned I venture out one last time, but this
time to hand water and see how the garden
looks in the shade of the evening. The birds are
quieter now and things begin to slow. I tuck in
the garden until we meet again tomorrow."*
— Sarah

What's Growing Outside in the Warm Season

Leaves
Arugula
Basil
Chives
Kale
Marjoram
Oregano
Rosemary
Sage
Swiss chard
Thyme

Roots
Ginger
Turmeric
Potatoes
Sweet potatoes

Fruit
Beans
Cucumber
Eggplants
Large gourds
Large melons
Okra
Peppers
Squash
Tomatillos
Tomatoes
Zucchini

Flowers
Calendula
  (if you miss
  cool season)
Cosmos
Coreopsis
Marigold
Nasturtium
Petunia
Strawflower
Zinnia

Warm Season at a Glance

MONTH 1: Planting the warm season garden, starting hot season plants indoors
>   Week 1: Plant large plants, start next season's plants indoors (hot or cool season).
>   Week 2: Plant small plants and seeds.
>   Week 3: Plant the garden border.
>   Week 4: Plant the interior of the garden.

MONTH 2: Tending warm season plants
>   Week 1: Feed plants.
>   Week 2: Support plants.
>   Week 3: Prune plants.
>   Week 4: Defend plants.

MONTH 3: Harvesting warm season plants
>   Week 1: Harvest herbs and greens.
>   Week 2: Harvest roots.
>   Week 3: Harvest small fruit.
>   Week 4: Harvest large fruit.

MONTHS

When a gardener really likes to pack plants into the garden the way I do, it may seem impossible to move between seasons without ripping everything up and starting afresh.

But not so fast—a slow transition is best.

Slowly switching from the cool season to the warm season in your garden keeps your soil covered, which means you don't have to water as often or worry about weeds sprouting up in bare soil. When you retain some cool season plants in the garden while you add warm season ones, your newest plants will have the support and insulation they need when they're at their most fragile state.

How do you decide which plants can stay and which should go? To find the keepers, look at the plant's harvest window. Cool season kale and Swiss chard, for instance, are biennials, which means they are genetically designed to stay in the garden into the warm season to complete their life cycle. Unless you experience a year when cool season jumps directly to hot, carrots will also need the rest of the summer to reach mature size.

Pull plants that have bolted (begun to go to seed), gotten tired, or struggle in warmer temperatures: leaf lettuces like 'Black Seeded Simpson', arugula, and romaine as well as root crops like radishes. Peas are likely finishing up as the warm season begins, although some varieties may need a few more weeks to finish fruiting.

With the spent or tired plants removed, it's time to set warm season plants between the cool season ones. For instance, when napa cabbages are nearly ready for harvest, I find a spot between each head and install a pepper plant. To avoid damaging existing plants, use a narrow hand tool like a hori hori to dig a compact but deep hole, then slip a pepper plant in among the cabbages. In the coming weeks, as you harvest the final cabbage heads, you can add more warm season plants to the recently vacated spaces. While these warm season seedlings are getting used to their new surroundings, they're well insulated by the surrounding older plants.

Once warm season plants are planted, it's important to harvest strategically. If a warm season plant is small and being shaded by a cool season plant, prioritize harvesting leaves from the larger plant. My kale, for instance, often blocks sunlight from my warm season eggplant, so I make sure to cut those kale leaves for my morning green smoothie so that the eggplant gets more sun. Pretty soon the eggplant will be tall enough that I no longer worry about it getting enough light.

Next up, prioritize trellis time. Plant tomatoes right next to sugar snap peas just as the peas are starting to clamber up their arch trellis. Since pea plants have small root structures, they don't need a lot of room underground, which allows you to dig a nice, deep hole nearby for your tomatoes. As soon as the pea plants stop producing, cut them from the base, tug the vines away from the trellis, and let tomatoes, cucumbers, or pole beans claim the trellis to do their thing.

As cool season lettuce plants and radishes finish and are harvested for the last time, plant pole cucumbers in the now open spaces. Take the hoops used during winter for frost cloth supports and swap them for obelisk trellises, which provide structure for the warm season plants as they grow and vine. Push the trellises at least 6 inches into the soil to keep them in place.

All told, it may take 3 to 4 weeks to fully transition your garden from the cool season to the warm one. By pulling spent plants every few days, you'll get to make the most of all that you planted in the cooler part of the year while also protecting your new warm season plants so they don't begin growing in a bare garden.

"I work from home. I take a few minutes on my lunch break to go outside and garden."

— @the_bluegrass_blacksheep

WARM SEASON/MONTH 1: PLANTING

There are exactly two modes when it comes to warm season gardening: "Plenty of time" and "Oh no!"

I have this conversation with nearly every gardener. All during the winter and colder parts of the year, they look at me and ask, "Isn't it too early to start?" And then as soon as the weather warms, they look at me again and say, "Isn't it too late now?"

It's as if there's this one magical day when you're allowed to start planting outdoors and you're either waiting on it or you missed it.

There's no need to fear—that day is more like a month, and even if you miss the first one, there's still time left in the second. These long, sunny days are forgiving, and they add up. It does benefit you to make the most of each one, but don't worry about being too late for the warm season: there's always something that you can grow and enjoy in the warm days of the year.

Warm Season/Month 1/Week 1

The first planting week in the warm season is perfect for hardening off seedlings. Hardening off could also be called *warming up*, as this process helps seedlings acclimate to the outside world before they are set permanently into the garden.

Hardening off is especially important for plants like tomatoes, peppers, and squash, which can be a bit sensitive to the big change. Here's how you can do it a few weeks before the last frost to make sure your plants are tough enough to handle the outdoors.

About 2 weeks before you plan to plant seedlings outside, start taking them outside a little bit each day. On the first day, let them hang out in a shady, protected spot for just a couple of hours. Avoid direct sunlight and strong winds, as this can be too much stress for them at first.

Each day thereafter, gradually increase the amount of time outdoors and gradually move them into partial sun and then to sunnier spots. By the end of the first week, the seedlings should spend most of the day outdoors. Continue to bring them inside at night, and keep an eye on the forecast. If the weather will be really cold or super windy, keep your seedlings indoors that day. They're toughening up, but they're not ready for the big leagues yet.

Water seedlings as usual while they're hardening off, but be mindful that exposure to sun and wind can dry soil. Provide additional water when needed, but don't overdo it. Soggy soil isn't good for young plants' roots.

After about 2 weeks hardening off, your seedlings should be ready to face the wind, sun, and temperatures waiting for them in the real wild world of your garden. Wait until your region's last frost date has passed, then officially set your warm season plants into the garden.

One more consideration this month: if you live in a hot climate, order or gather hot season seeds now, because you'll need to start the following seeds indoors soon to be ready for the planting out in the hot season.

Leaves
Arugula
Basil
Malabar spinach
Mustards
New Zealand spinach

Fruit
Armenian cucumber
Crowder peas
Hot pepper
Small eggplant
Squash
Tomatillo
Yard-long bean
Zucchini

If you live in a mild climate and your hottest season is the warm season, then your second season (see page 243) will be the cool season. So it's time to order and start the following seeds soon for your second cool season.

Leaves
Buttercrunch
Cabbage
Cilantro
Dill
Fennel
Iceberg lettuce
Kale
Parsley
Romaine
Spinach
Spring mix
Swiss chard

Roots
Beets
Carrot
Radish

Fruit
Broccoli
Brussels sprouts
Cauliflower

Flowers
Calendula
Nasturtium
Pansy
Viola

"As a substitute teacher, swim coach, and mom of two, I use the garden to slow down and relax. The output is just a bonus."

— Ally

Warm Season/Month 1/Week 2

This week, you'll begin making way for the large warm season plants.

First, remove bolted or spent plants from your garden. Focus on the leafy green plants first: arugula, romaine, buttercrunch, radishes, cilantro, dill, mustards, bok choy, and spinach. They're most likely finished growing by this point in the season, and you can simply cut the plants at soil level if their root systems won't interfere with what you'll plant next. As long as the plants were healthy and disease-free, you can add them to your compost pile to break down and eventually contribute to your garden's health.

Then harvest the mature radishes, carrots, and beets.

To these new blank spots in the garden, add a few inches of organic compost. Warm season plants, especially the fruiting plants, will need a fresh supply of nitrogen, phosphorus, potassium, and micronutrients. Adding soil amendments now will support your plants' growth for many months to come.

gardener time

"On work-from-home days, my garden is my commute. I go out when my husband comes home."

— @saragriffinrdn

If your cool season garden wasn't as productive as you'd hoped, take a minute to do a soil test (see page 86) to identify any nutrient deficiencies. The test results will recommend ways to boost the soil health, whether by incorporating bonemeal to improve phosphorus levels or blood meal to provide nitrogen. Mix the amendments into the soil and water the garden bed well before adding the next season's plants.

Now it's time to plant! Prioritize the largest plants first, as these need more growing days than smaller plants. Start with tomatoes, peppers, eggplant, squash, zucchini, cucumber, and melons.

Set vining tomato plants along your tallest trellises. Place eggplants and peppers in front of the tomatoes, using stakes or cages to support them. Site zucchini and squash in 1-square-foot spaces or alongside the edges of the garden to make the most of your space. Add cucumber and pole bean seeds or young seedlings near the tallest trellises so they have support as they grow to full maturity.

Give Me Five

Got a free moment or two? Here are garden tasks you can do in just 5 minutes.

- Remove bolted lettuces from the garden.
- Pull up bolted dill and cilantro (but leave some for the bees!).
- Plant tomatoes along your trellises.
- Plant eggplants, squash, and zucchini.
- Add peppers in front of the tomatoes and eggplants.
- Make a big spring salad to celebrate the end of the cool season.
- Pass along extra lettuces to friends and neighbors.

Warm Season/Month 1/Week 3

This week we add smaller plants in the newly opened areas.

Bush beans, arugula, and basil are great plants to place around larger ones, as they have smaller root systems and don't take up much space. Bush beans fix (capture) nitrogen from the soil and deliver it to the roots of nearby plants, which is beneficial for heavy feeders like tomatoes. Arugula, with its shallow root system and quick growth habit, makes an ideal filler between slower growing larger plants. Basil also grows well alongside the larger plants of the warm season garden.

When planting seeds around larger plants, consider the spacing. Sow bush bean seeds 3 to 4 inches from tomato or pepper plants so the beans have enough room to grow without interfering with the larger plants. Arugula seeds can be sprinkled lightly in spaces between other plants, as these greens can be grown densely and harvested young.

This is a staggered planting approach that gives tomatoes and peppers time to establish themselves before the smaller plants fill in the gaps to create a lush, productive garden bed.

"I picked up gardening a year and half ago, right before my son was born. When baby goes down for a nap, I give myself time to decompress in the garden."

— Hannah

Give Me Five

If you have 5 minutes, here are a few things you can get done in the garden:

- Check on plants installed last week.
- Plant bean seeds around newly installed plants.
- Add arugula seeds in open spots.
- Add basil plants near pepper plants.
- Water in new plants well.
- Harvest herbs to enjoy over the weekend.
- Make an herb dipping sauce for dinner.

Warm Season/Month 1/Week 4

Last but not least, more herbs and flowers! Herbs and flowers are the epitome of multifunctional. They're beautiful, fragrant, and often edible while attracting beneficial insects like bees and butterflies and warding off some unwanted bugs.

Basil, a summer culinary favorite, thrives in the warmer months and is a great companion for tomatoes, peppers, cucumbers, and squash. Thanks to its strong scent, basil can help repel pests like mosquitoes and certain flies. This herb requires plenty of sunlight and regular watering, so place it in a sunny spot where it can bask in the warmth of the summer sun.

Marigolds aren't just pretty to look at—their strong scent acts as a natural pest deterrent for nematodes, aphids, and other pests. Plant marigolds around the borders of your vegetable garden.

Zinnias, cosmos, and coreopsis attract pollinators like bees and butterflies to support the health of your vegetable plants. They come in many vibrant colors, and the plants can get quite tall, making them perfect for growing along the borders of your garden rather than inside the planting beds. As a bonus, they're relatively easy to grow and maintain.

Because of their sunlight, space, and soil requirements, most annual herbs and flowers are best sited along the borders of your garden. Intersperse different types for a mix of colors and heights. Water them well for the first 2 weeks; after that, these plants should thrive.

By the end of this month, your 5 minutes a day have added up to 150 minutes of preparing and planting for the warm season. You began with a garden full of lettuce, peas, and cabbage, and you wrap up with the promise of tomatoes, cucumbers, beans, squash, basil, and so many flowers.

The warm season garden may still resemble the cool season, but hidden underneath those greens of last season is the promise of an overflowing and abundant fruiting garden for months to come. And if you enjoyed the season's lengthening days and the warmer air, you're going to *love* what comes next.

WARM SEASON/MONTH 2: TENDING

It's tending month: time to keep it all going, to take care of the seeds and plants that are now 3 or 4 weeks old in the garden. Don't blink. These plants are changing by the minute, and they'll need your attention with each new inch.

First, set a schedule to feed your plants. Be sure the warm season plants get plenty of water and nutrients to make it through their "teenage weeks" as they grow from infant to adult.

Next, focus on support, ensuring your plants have the structures and root system needed to grow to their full potential.

Then, rein in that growth through pruning so that you get more of what you want and less of what you don't.

Finally, focus on protection. In this week, you'll be certain that you get to your harvests before pests or diseases do.

Even though you're dealing with water, pruning, and pests, don't stress. These 5-minute gardener practices let your organic garden (mostly) take care of itself.

By planting new plants each season, you disrupt pest life cycles because different plants are grown in the same spot every 2 to 3 months—a sort of mini crop-rotation schedule.

By adding compost and other soil amendments each time you replace plants, you build a nutrient-rich base for stronger, healthier plants more capable of fighting off pests and disease.

And since you've planted herbs and flowers along the border of each bed, you've created a diverse garden that deters pests and attracts beneficial insects. Finally, by harvesting from the garden often, you ensure that no plant stays in the garden too long to attract pests or disease to your healthy garden.

In other words, you can do this.

Give Me Five

Got 5 minutes? Here's how to make the most of those moments this week:

- Make a schedule to care for your plants.
- Water in your newly installed plants.
- Prune off any spoiled leaves from new plants.
- Place trellises or supports alongside new peppers and eggplants.
- Water in newly planted seeds.
- Harvest herbs and basil.
- Make pesto pasta and serve with fresh arugula salad.

Warm Season/Month 2/Week 1

Some people refer to tomatoes, peppers, eggplants, and squash as "heavy feeders," but I think it's simpler to just understand that they're large and long-season plants. Yes, they need a lot of water and nutrients, but that's in part because they're in the garden for many months and they have so long to go before they begin to produce fruit.

So, yes, you could call them "needy," but more to the point, you could call them "productive."

Water is crucial for these plants, especially as they first get established, and then as they set fruit and grow. These plants prefer a consistent watering schedule, and it's best to water deeply a few times a week rather than a little bit every day.

Use a 5-minute spot for deep watering, which encourages the roots to grow deeper into the soil, making the plants more drought tolerant and sturdy. However, be careful not to overwater, as soggy soil can lead to root rot and other diseases. When the top inch of soil feels dry to the touch, choose a morning or evening for 5 minutes of a deep water session.

In the first warm season month, you provided a rich soil for these plants to grow in. But at each new stage of growth, it's important to add nutrients to the soil to support the plant as it flowers, puts on fruit, and ripens. Plants need nitrogen to produce leaves, potassium to grow strong roots, and phosphorus to fruit. So choose an organic amendment that provides the nutrients necessary at each stage of growth.

Once watered and fed, be certain these warm season plants receive plenty of sun: at least 8 hours of direct sunlight each day. The more sunlight they get, the better they will produce. Pull or prune cool season plants or greens like kale, Swiss chard, or herbs that stand in the way of these plants' sunlight needs.

"This is how my toddler and I fill our time between getting home and bedtime. We garden together."

— Megan

Warm Season/Month 2/Week 2

All that nutrient-rich soil will result in so much plant growth this month. As stems, vines, and leaves expand and stretch further by the day, give each plant something to lean on.

For vining tomatoes, pole beans, and cucumbers, this means installing a trellis. Head to the garden this week with twine, scissors, and a compost bag. Spend a minute with each vine, tying the leading vine stem to the trellis and being careful to not let stems cross one another.

Bushy plants like peppers, eggplants, and squash can be supported with stakes or small tomato cages.

Build an Arch Trellis

For a sturdy way to support vining plants like tomatoes, pole beans, or cucumbers, build an arch trellis using a metal piece of cattle panel fencing and four rebar stakes you can purchase at a local farm supply store. You'll need a helper to create the arch. If your cattle panel is wider or narrower than the one specified here, adjust the rebar locations as needed. If possible, make the arch 4 feet wide and at least 6 feet tall.

Hammer or mallet

Four 2-foot-long pieces of rebar

One 16-foot-long, 50-inch-wide cattle panel

Zip ties or garden wire

Wire cutters

First, select a location in your garden that gets enough sunlight for your vining plants and has ample space for the arch. Level the ground to ensure stability.

Using the hammer, drive the pieces of rebar about a foot into the ground, leaving enough above ground to support the cattle panel. Space the rebar 48 inches apart so there's plenty of height and width to walk and push a wheelbarrow through the arch opening.

With someone's help, lift the cattle panel and bend it into an arch shape. Position one end of the cattle panel over two of the rebar stakes on one side of your intended arch location. Then carefully bend the panel and guide the other end over the corresponding rebar stakes on the other side.

Once the cattle panel is in place over the rebar stakes, use zip ties or garden wire to secure the panel to the rebar. Make sure the panel is tightly secured at several points along each stake.

To avoid potential injuries while working around the trellis, use wire cutters to trim off any sharp ends or excess material from the zip ties or wire.

With the trellis installed, attach vining plants like tomatoes, pole beans, or cucumbers at the base of each side of the arch. As the vines grow, gently guide them onto the trellis and secure loosely with garden twine every 6 to 12 inches. Regularly check the stability of the trellis, especially after strong winds or heavy rain.

Give Me Five

Got 5 minutes? You can do the following things this week:

- Gather materials to build a trellis.
- Water in your plants.
- Build a new trellis.
- Put stakes around your peppers.
- Hill the beans in the garden.
- Harvest herbs and arugula.
- Make vegetable soup with broth from vegetable stems and garden herbs.

Warm Season/Month 2/Week 3

The time has come when you can no longer go to the garden empty-handed. Pruners are now the entry fee.

If you haven't been pruning your plants yet, it's time to start. Pruning vining plants like cucumbers, pole beans, and tomatoes is an essential part of garden tending, especially if you follow my intensive planting method. Proper pruning helps to manage plant size, improve air circulation, and direct the plants' energy toward producing fruit rather than excess foliage.

To prune cucumbers, first remove any yellow or dead leaves as well as any cucumbers that look misshapen or damaged. Identify the main stem and any lateral (side) shoots. As the plant grows, you can prune away some of the lower lateral shoots to direct the plant's energy on the main vine. This encourages the plant to produce more cucumbers while also improving the air circulation to prevent disease. Never prune more than a third of the plant's leaves within a week's time.

To prune vining tomatoes, first remove the oldest and lower leaves to help prevent soilborne diseases from splashing up onto the plant during watering or storms. Then it's time to decide whether you'll keep or prune the suckers (the new tomato shoots) that grow in the axils between the branches of the main stem. These suckers can eventually grow into full-size branches, but removing them helps the plant direct more energy into fruit production on the main stem.

I typically leave the suckers once the plant grows past 3 or 4 feet tall. Pruning the suckers results in less fruit, but the fruit is bigger. Leaving the suckers and trimming back the nonproducing stems results in more fruit that's smaller. If you choose to prune the suckers, cut the stems off when they are small, preferably less than 4 inches long.

To prune nonvining tomatoes, remove only the lower and spent leaves, and let everything else keep growing as it wants.

If you prune your tomato plants once a week, you'll encourage stronger, healthier growth and a more abundant fruit yield. Not only does this practice increase the quality of your harvest, but it also keeps your garden looking tidy and well maintained.

Pole beans require less pruning than tomatoes or cucumbers. These plants naturally climb and produce fruit along their vines. However, you can prune away any leaves or stems that look unhealthy or are overcrowding other plants. Removing lower leaves can also improve air circulation around the base of the plant, reducing the likelihood of fungal diseases.

General pruning tips:

- Always use clean, sharp pruning shears to make precise cuts. This helps prevent the spread of disease.

- It's best to prune in the morning when plants are well hydrated.

- Regularly check your plants for pruning needs, as growth can be rapid, especially in the peak of summer.

- Be mindful of the plant's overall health and avoid removing too many vines or leaves, which can stress the plant.

I think it was the famous sculptor Michelangelo who said, "The sculpture is already complete within the marble block before I start my work. It is already there. I just have to chisel away the superfluous material." As a gardener, you are like a sculptor, and your pruners are your chisel. You chisel away the extra or superfluous plant material to showcase the masterpiece growing underneath.

"Gardening is my postwork meditative practice.
Harvest, water, take a pic, and breathe."

— @sbalsitis_5

Warm Season/Month 2/Week 4

By this week, it's clear that fruit is on its way. And since you're growing an organic garden, the bugs, squirrels, and other creatures are on their way too.

It's time to protect your garden.

Let's start with the bad news. Tomatoes are most often attacked by aphids, hornworms, and whiteflies. Aphids are small, sap-sucking insects that can weaken plants and spread diseases. Hornworms are large green caterpillars that can quickly defoliate a tomato plant. Whiteflies are tiny white insects that feed on plant sap and can cause the leaves to yellow.

Beans can attract pests like Mexican bean beetles and aphids. The larvae and adults of Mexican bean beetles can skeletonize bean leaves, severely harming the plant's health. Aphids can weaken the plant and spread diseases.

Cucumber beetles and spider mites are walking their way to your cucumbers. They not only eat the leaves and fruit but can transmit a disease known as bacterial wilt. Spider mites, while tiny, can cause significant damage by sucking plant juices.

Squash and zucchini are susceptible to squash bugs, squash vine borers, and cucumber beetles. Squash bugs can suck sap from the leaves, causing them to wilt and die. But much worse, squash vine borers, which burrow into stems, can kill plants.

Ready for some good news? You've already taken steps to protect your garden.

By planting herbs and flowers around your garden perimeter, you provide food and habitat for ladybugs, parasitic wasps, and other beneficial insects that prey on these pests.

By changing the plants each season, you ensure no pest has the same food in the same spot for a long period of time.

By spending 5 minutes in the garden each day, you can see the first sign of pest or disease damage and can stop it before things get worse.

When pests are present, prune away affected leaves and stems. Next, clear the soil area under the plant to remove any debris and insect hiding spots. Spread some compost around the base of the plant to support the plant with a boost of nutrients as it fends off its predators. At the next sign of flies or pests, spritz the plant with garlic spray or castile soap mixed with water on leaves that are affected (see page 98). Check on the plant each day for the next week, repeating this process and removing pests each time. If after three or four applications the pest issue does not resolve, it's best to pull out the affected plant and move on.

Give Me Five

It only takes 5 minutes to get the following tasks done this week:

- Gather materials to protect your garden.
- Prune away lower leaves.
- Add compost around the base of plants.
- Make homemade garlic spray.
- Use garlic spray on affected leaves.
- Harvest the first beans and herbs.
- Make garlic green beans, herbal sauce, and rice.

WARM SEASON/MONTH 3: HARVESTING

I'm not a betting woman, but if I were, I'd bet you pictured yourself holding a basketful of tomatoes, cucumbers, and watermelon once or twice, or a million times, when you thought about becoming a gardener.

Well, through the cold season and the endless days of gathering and planting seeds, you may have wondered if that day was just wishful thinking. But this month, you'll find it's not just a wish, it's a reality. Because it's finally time to harvest in the warm season.

Warm Season/Month 3/Week 1

The sun is up, the days are long, and the edges of your garden beds are covered in delicious and fragrant herbs. It's finally time to harvest as many green herbs as possible, both to enjoy now and to save for later.

Use clean, sharp tools. And if you want the best flavor, harvest in the morning after the dew has dried but before the sun gets too hot. This is when the herbs' oils are most concentrated.

Start with rosemary's woody stems and needlelike leaves. Harvest the softer, newer growth at the tips of the branches. This part of the plant is more flavorful and tender, perfect for cooking. Remember not to take more than one-third of the plant at a time so as not to stress the plant.

"I'm a single mom with two jobs, so my garden time is precious and necessary for my sanity."

– Barbara

Moving on to thyme, a more delicate herb with tiny leaves on thin stems. The best way to harvest thyme is by snipping the stems just above a growth node or a set of leaves. This encourages the plant to branch out and grow more densely. You can usually collect a good amount of thyme without making a big impact on the plant's overall health.

Sage is next. With its larger leaves, sage is easy to harvest but slower growing, so it's important not to over harvest. You want to cut whole stems, ideally those that are showing new growth. Cut close to the base of the plant, but leave a few inches of stem with leaves on the plant to ensure it keeps growing.

Last but not least, there's oregano, the most sprawling of the herbs. Harvest oregano by snipping the stems, just like thyme. Aim to cut just above a leaf node or pair of leaves. Oregano can handle a more aggressive harvest compared to the other herbs, but still, it's good practice to leave enough foliage so the plant can continue to grow more.

Harvest basil as soon as the plant has a good number of leaves, typically 6 to 8 weeks after planting. Snip off the top leaves just above a pair of larger leaves. This method encourages the plant to branch out and produce more leaves. Frequent harvesting keeps the plant from flowering, which can alter the flavor of the leaves, plus you'll gain a steady supply of fresh basil for your kitchen.

Beyond herbs, this is the week to start harvesting tender warm season greens like mustards, mizuna, arugula, and collards. These leafy greens, each with their unique flavor profiles and textures, are not only delicious but also packed with nutrients.

Mizuna, with its feathery, delicate leaves, is easy to harvest. This Japanese green is known for its mild flavor and can be harvested in a similar way to mustard greens. Snip or pinch off the outer leaves, ensuring that you leave the central part of the plant intact. Mizuna grows back quickly, so regular harvesting encourages more growth.

Arugula is my favorite for warm season salads and sandwiches. The younger the leaves, the milder the flavor. You can start harvesting when the leaves are 2 to 3 inches long. Cut the outer leaves at the base, just above the soil line. Avoid picking the center leaves, as this is where new growth emerges. Each plant will provide at least two, if not three, harvests like this one.

Collect mustard leaves when they are young and tender, as older leaves can become quite tough and overly spicy (ask me how I know). Start harvesting when the leaves are 4 to 6 inches long. Simply use your fingers or a pair of scissors to snip off the outer leaves, leaving the inner leaves to keep growing.

Lastly, collards, known for their large, hearty leaves, are a staple in many cuisines. Wait until the plant is well established and the leaves are about the size of your hand or larger, then harvest the lower, older leaves first, cutting them off at the stem. This approach encourages the plant to keep growing upward and producing more leaves.

Always harvest greens in the morning when the leaves are most crisp, and make sure to wash them thoroughly before using. Regular harvesting not only keeps the plants healthy but also ensures a steady supply of fresh, tasty greens for summertime meals.

Give Me Five

Here are a few tasks you can do in just 5 minutes this week:

- Pick your favorite recipes to make with fresh garden herbs.

- Harvest oregano, sage, and thyme.

- Hang half of the oregano, sage, and thyme harvest to dry.

- Harvest basil.

- Freeze basil with olive oil for future use.

- Harvest greens and mustards.

- Make a quick omelet with garden greens.

Warm Season/Month 3/Week 2

Week 2 is a deep dive underground—into roots, that is. Harvesting warm season root crops like garlic, potatoes, parsnips, and turnips is a little more of a challenge because your harvest isn't in plain sight. These delicious foods seem to be hiding beneath the soil's surface. But if you look closely, each root crop has its own sign and timing for the perfect harvest.

Harvest these crops on a dry day, as wet soil can make the process messier and more difficult. And don't forget to be gentle. Root crops can be damaged during the harvesting process, which can lead to spoilage.

Harvest potatoes first. You'll know they're ready when the plant's foliage starts to yellow and die back. For new potatoes, harvest the tubers when the plants are still flowering. Starting at the base of the plant, gently dig with a fork or your hands, being careful not to stab the potatoes. You might need to dig in an area a little wider than the row to find all the tubers, which can spread out under the soil.

Garlic is usually ready for harvest 90 to 100 days after the last frost of the season, when the lower leaves start to brown. Gently loosen the soil around the bulb with a spade or garden fork, being careful not to damage the garlic (it bruises easily). Once the soil is loosened, lift the bulbs right out of the ground. Brush off clumps of soil, but avoid washing the bulbs until you're ready to use them. Hang the garlic to cure in a dry, ventilated space for a few weeks so the skins harden and then start to enjoy the most delicious flavor that you grew yourself.

Warm Season/Month 3/Week 3

You've picked your share of green things from the garden this month, but it might finally be time to pick something red, yellow, orange, and even purple.

Bush beans are one of the stars of the quick-harvest garden, usually ready around 60 days after planting. The key to harvesting bush beans is to pick them when they are firm and crisp but before they become too large and tough. The beans should be a few inches long, slender, and snap easily when bent. Regular picking is essential, as it encourages the plant to produce more beans. To avoid damaging the plant, use two hands to pick—one to hold the plant and the other to gently pluck the bean.

Cucumbers are ready to harvest when they are firm, green, and usually 6 to 8 inches long for slicing varieties, smaller for pickling types. Overripe cucumbers can become bitter and seedy, so timely picking is crucial. Gently twist or cut the cucumber off the vine, being careful not to damage the plant. Frequent checks are a good idea, as cucumbers grow quickly in the right conditions.

Summer squash and zucchini are typically ready for harvest around 60 days after planting. These vegetables are best when they are young and tender, usually 6 to 8 inches long. They should feel firm and have glossy skin. Like cucumbers, they should be cut or gently twisted off the vine. If left to grow too large, the fruit can become tough and less flavorful. Regular harvesting encourages the plant to produce more fruit.

Cherry tomatoes are typically ready when they've reached their full color, be it red, yellow, orange, or even purple, depending on the variety. The best indicator of ripeness is color and a slight give when gently squeezed. To harvest, gently hold the tomato and twist it off the vine, or use a pair of scissors to snip it off, leaving a bit of the stem attached. Regular harvesting encourages the plant to produce more fruit, so don't be shy about picking them.

Pole beans are another prolific producer in the garden. The key to harvesting pole beans is timing—pick them when they are firm and before you can see the seeds bulging inside the pods. This ensures they are tender and not stringy. Gently pull or snap the beans off the plant, being careful not to damage the plant. Regular picking stimulates more production.

Eggplants are ready to harvest when their skin is glossy and the fruit is firm. The size can vary depending on the variety, but generally, eggplants should be picked before they start to lose their shine and become dull, which indicates overripeness. To harvest, cut the eggplant from the plant with pruning shears, leaving about an inch of the stem attached. It's important not to pull or twist eggplants off the vine, as this can damage the plant.

Harvest in the morning when temperatures are cooler to be sure all your fruit stays crisp and doesn't wilt.

Give Me Five

If you have 5 minutes this week, you can get all these tasks done:

- Choose your favorite recipes to make with fresh garden harvests.
- Pick beans in the early morning.
- Store beans for future use.
- Harvest the first cucumbers and squash.
- Parcook and store extra squash in the freezer.
- Harvest the first small peppers.
- Make a dish with squash, beans, and peppers.

Warm Season/Month 3/Week 4

By this point, your harvest basket is packed with herbs and greens, potatoes, garlic, peppers, squash, and even a few tomatoes. But wait, there's more!

If there's been enough rain, enough sun, and enough care, the larger fruit in your garden may be ready to pick too.

To harvest watermelons, look for a few key signs of ripeness. The first is the color of the bottom spot where the melon rests on the ground; it should change from white to a creamy yellow. Also, the tendrils near the stem of the fruit will dry and turn brown when the watermelon is ripe. But wait—before you cut the stem, give the watermelon a tap; a ripe melon will have a hollow sound. When cutting the watermelon from the vine, use a sharp knife or shears, leaving a few inches of stem attached to prevent the fruit from rotting.

Cantaloupes, or muskmelons, tell you they're ripe by detaching easily from the vine. This is known as slipping. You should also see the skin beneath the netlike texture change to a golden or yellowish color, and the fruit should emit a sweet, musky aroma (the name fits!). Once you notice the slipping stage, cut the fruit from the vine, leaving a small portion of stem attached.

Harvesting large gourds, such as bottle gourds or birdhouse gourds, requires waiting until the stem near the fruit begins to dry and turn brown. The shell of the gourd should be hard; you can test this by attempting to press your fingernail into it. If the shell is hard and the stem is brown, cut the gourd from the vine with a few inches of stem remaining. These gourds often require a curing process postharvest to ensure they dry properly.

WEEKS

During the season of cold days, it can be hard to motivate yourself to get outside and put 5 minutes a day toward the garden. But now the challenge is likely more about gardening within your allotted 5 minutes. When the weather is warm, the sun is shining, and there's promise of rain, it can be hard to stop after just 5 minutes. But here's a plan you can follow each week to make the most of your moments, even if you wish you could stay longer.

"The garden is the first place I go when I get home from work to de-stress."

— Aly

Day 1

There's an old saying that goes, "Give me 6 hours to chop down a tree and I will spend the first 4 sharpening the ax." I can't tell you who said those words, and I can't tell you how to chop down a tree. But I can tell you that making a plan and preparing for the week ahead in the garden is the key to the rest of the week's tasks feeling possible.

Start the week by checking the weather forecast and making a plan to fit your 5 minutes in where it counts most.

First, check for the warmest days this week. By anticipating hot, dry days, you can adjust your watering schedule to ensure plants receive adequate hydration without overwatering, which can be just as harmful. But if rain is expected, you might reduce or skip watering to prevent waterlogged soil and to conserve water.

If a day is expected to be very warm and humid, make time to be on the lookout for fungi and more pests. Do not apply sprays to the leaves before a predicted increase in humidity or temperature, which can only make matters worse.

On the hottest days, make time to observe your plants and be sure they're not stressed.

Based on the forecast, pick the coolest parts of the day, typically morning or late afternoon, for planting and harvesting. Plan tending tasks like pruning and supporting any time of day when you're able to get outside.

Even with all the planning, as financial analyst Patrick L. Young said, "The trouble with weather forecasting is that it's right too often for us to ignore it and wrong too often for us to rely on it."

You and I both know that there's a 50-50 chance the forecast will be wrong. But with a plan in place, you have a starting point and spots in each day's calendar that ensure you'll make your 5 minutes happen.

Day 2

Add large plants to the garden on this day. Tomatoes, cucumbers, squash, and zucchini can all be planted twice within a season if it's long enough.

Since these plants generally take between 60 to 80 days to mature from transplant, the first harvest often occurs in midsummer.

If your area has an extended warm season, you can implement a staggered planting approach for continuous harvests. By starting a second round of plants indoors and timing their transplant into the garden, you can ensure a steady supply of fruit harvests almost every day of the season.

The second batch of large plants can be transplanted outdoors as the first batch nears the end of its productive period. These fruiting plants tend to produce prolifically but may wear out or become more susceptible to pests and diseases as the season progresses. This staggered approach ensures that as the first set of plants begin to decline, often due to accumulated stresses or diseases, the second set is just coming into its prime, ready for harvest.

By planning a second planting approximately 60 days into the warm season, you can rejuvenate the garden's productivity. This second wave of plants will reach peak production as the earlier plants are winding down so the harvests never stop.

You can use the second planting to experiment with different varieties of cucumbers. Plant an early-maturing variety of cucumber initially, then choose a different variety for the second planting to try out different flavors and growth habits within the same season.

Staggered planting also helps with managing pests and diseases. By introducing new plants midseason, the risk of certain pest infestations and diseases can be mitigated, as these issues often build up over time. Fresh plants are less likely to be affected immediately, giving them a chance to establish and produce a good yield before these challenges become significant.

To make staggered planting work, you've got to be mindful of the timing and the specific needs of each plant variety as well as understand the length of the growing season and the maturation time of each crop.

Use the second day of each week to focus on the large plants for this season, a possible second planting as well as the season to come, and your garden will always be full of large and productive plants.

Day 3

Today's the day to add new small plants throughout the garden in the spaces that continue to become available as early-season plants complete their life cycle and bolt (go to seed).

When bolting occurs, leaves often become bitter, and the plant focuses its energy on seed production. Once bolting occurs, the culinary quality of these herbs and greens diminishes. When you see signs of bolting, remove the cool season plants to clear space in the garden and prevent these plants from self-seeding.

Cool season cilantro, dill, and arugula can be followed by warm season plants such as peppers, basil, and beans. When lettuces, radishes, and carrots are pulled from the garden, replace these with marigolds, zinnias, arugula, and mizuna.

This constant rotation in the garden allows for a diverse range of plants, beneficial for soil health and pest management. Different plants have varying nutrient requirements and pest associations, so changing crops helps break pest and disease cycles and prevent soil nutrient depletion.

Day 4

Maintaining a clean and well-tended garden during the warm season is vital for the health and productivity of your plants and can be achieved just 5 minutes at a time. Use a small garden rake to clear debris and fallen leaves from underneath plants, and cut spent plants at the soil level. Garden debris, such as fallen leaves, rotting fruit, and dead plants, are ideal hiding spots and breeding grounds for insects, slugs, and disease-causing organisms.

Decaying plant matter can attract aphids, snails, and cucumber beetles, which not only feed on plants but can also spread diseases. By regularly removing this debris, you reduce the potential for these pests to establish themselves in your garden.

Also, as plant materials decompose, they can consume soil nitrogen, temporarily making this crucial nutrient unavailable to your plants—particularly problematic during the growing season when plants are actively developing and require ample nutrients. While decomposition is a natural and beneficial process, contributing to soil health in the long run, excessive and unmanaged debris can lead to nutrient imbalances in the short term. It's better to let the plants decay and become compost elsewhere and then place the finished compost back into the garden once it's finished.

Every few weeks on this day, after clearing the garden of debris, add a shallow layer of compost or sprinkle earthworm castings over the soil. This is my favorite way to mulch the garden. A fresh layer of compost maintains soil moisture, which reduces evaporation and keeps the soil cooler, providing a more stable growing environment for plant roots. Additionally, compost and earthworm castings are rich in nutrients and beneficial microorganisms. As they break down, they slowly release nutrients back into the soil, improving its fertility.

These organic materials also enhance soil structure, promoting better aeration and drainage, which are essential for healthy root development and can also act as a physical barrier, deterring some pests and reducing the likelihood of soilborne diseases coming into contact with plant foliage.

This 5-minute activity of clearing debris and adding a little compost keeps pests from eating your food while giving your plants the boost they need to grow to the next level.

Day 5

Pruning is critical for managing an intensively planted warm season garden, particularly in the case of indeterminate, or vining, plants that can grow 6, 7, even 10 feet long, such as tomatoes, cucumbers, and certain types of squash and beans. These plants benefit significantly from regular pruning, as it helps direct their energy toward fruit production rather than excessive leaf growth, leading to a healthier plant and a more bountiful harvest.

Indeterminate plants, unlike their determinate counterparts, continue to grow and produce fruit throughout the growing season. However, without pruning, they can become overgrown with leaves and stems, which can lead to several issues. Too much foliage can reduce air circulation around the plant, increasing the likelihood of fungal diseases, and it can result in a dense leaf canopy that shades the fruit, hindering proper ripening. By pruning, gardeners can balance the growth of leaves and fruit, ensuring that the plant's energy is efficiently used.

Here's the best way to do it: First, prune away damaged or diseased leaves, as these can attract pests and spread diseases to other parts of the plant and garden. Then focus on older and lower leaves, especially those that are touching the ground, as these are more susceptible to soilborne pathogens and pests. Removing these leaves improves air circulation and reduces the risk of disease. Finally, remove unwanted stems or non-fruit-producing stems. If left unchecked, these can lead to a plant full of leaves and very little fruit.

Cutting away extra leaves also ensures that developing fruit receive adequate sunlight and air, promoting healthy growth and ripening.

But don't cut too much. Leaves are vital for photosynthesis and overall plant health, so be careful not to overprune. A good rule of thumb is not to remove more than one-third of a plant's leaves in a month.

Starting with damaged leaves, progressing to older and lower foliage, and then judiciously thinning out stems and excess leaves can significantly enhance plant health and yield.

Using Trap Crops

Using trap crops is a clever and environmentally friendly strategy for managing pests in the garden. A trap crop is a plant that's more attractive to pests than your main crop. When planted strategically, these sacrificial plants can divert pests, reducing the damage to your main garden. The key to success with trap crops is knowing which pests you're dealing with and what plants they prefer. For example, nasturtiums can attract aphids away from vegetables, marigolds can lure nematodes away from tomatoes, and mustard plants can draw flea beetles away from brassicas like cabbage and kale.

Before planting any garden vegetables, research the common pests in your area and their preferred plants. Then, plan your garden layout. Plant trap crops around the perimeter of your garden or between rows of your main crops. This placement is crucial—a trap crop should be close enough to lure pests but not so close that pests easily move to your main crops.

Timing the planting of your trap crops is as important as their placement. Ideally, you want the trap crops to be slightly more mature than your main crops, as pests are usually drawn to more developed plants. This might mean planting your trap crops a few weeks before your main crops.

Regularly check your trap crops for pest infestation. Once they are heavily infested, you have a few options. You can remove and destroy these plants, thereby removing a large population of the pests. Alternatively, you can treat the trap crops with an organic pesticide, which can be more targeted and less extensive than treating your entire garden.

For greater success, rotate your trap crops each season, just as you would your main crops. This practice prevents the buildup of pest populations in the soil and keeps your pest-management strategy effective.

Trap cropping works best as part of a broader integrated pest-management strategy. Combine this method with other organic practices, such as attracting beneficial insects, using row covers, and practicing good garden hygiene, to maximize your garden's health and productivity.

By incorporating trap crops into your garden, you can avoid the use of chemical pesticides and foster a more natural, balanced ecosystem. This approach not only helps in managing pests but also adds diversity to your garden, making it more resilient and sustainable.

Day 6

Tomatoes are best harvested when they are uniformly colored and slightly soft to the touch. Gently twist the fruit off the vine or use a pair of clean, sharp scissors or pruners. It's important not to pull the fruit harshly, as this can damage the plant. Wash tomatoes gently under running water and store at room temperature away from direct sunlight. Refrigeration is not recommended for tomatoes; it can diminish flavor and texture.

Harvest cucumbers when they are firm and bright green, before they start to yellow. Cut the cucumber from the vine with scissors or a knife, leaving a small portion of the stem attached. Wash cucumbers under cool running water and dry them thoroughly. They can be stored in the refrigerator's crisper drawer, ideally in a plastic bag to retain moisture.

Beans are best when tender and firm, before the seeds inside have fully developed. Harvest them by holding the stem with one hand and gently pulling the bean off with the other. Rinse beans in cool water and pat dry. They can be stored in a plastic bag in the refrigerator, where they will keep for about a week.

Squash and zucchini should be harvested when they are still small and tender, as larger ones can be tough and less flavorful. Cut the squash or zucchini from the vine with a sharp knife, leaving a small piece of stem attached. Wash gently under running water and store in the refrigerator, ideally in a vegetable drawer.

Peppers can be harvested at various stages, depending on the desired ripeness and flavor. Use a knife or scissors to cut the pepper from the plant, leaving a short amount of stem attached. Wash peppers under running water, dry, and store them in the refrigerator. They can be kept in a plastic bag in the vegetable drawer.

Eggplants are ready when their skin is glossy and they are firm to the touch. Harvest by cutting the stem with a knife, leaving an inch of stem attached. Rinse under water, dry, and store in the refrigerator.

After harvesting, washing the produce is important to remove any soil or residue. However, it's crucial to dry the fruits and vegetables thoroughly before storage to prevent mold and rot.

Most warm season produce is best stored in the refrigerator, except for tomatoes, which should be kept at room temperature.

It's also a good practice to use or preserve the harvest as soon as possible to enjoy the freshest flavors and highest nutritional value. You did all that work to grow seriously fresh food, so don't waste it!

Harvesting, Storing, and Preserving Herbs

Keep these basics in mind when you're harvesting and preparing your garden-fresh herbs to store and save.

Harvest herbs first thing in the morning, when their essential oils are most potent. Harvest the lower outer leaves first. Cut the herbs before they flower, unless your goal is to save seeds (like fennel seed or dill seed). Once a plant flowers, its focus is on producing seeds, not growing more leaves.

Inspect the harvest and shake or pluck out insects, debris, diseased or dead leaves, and so on.

There are two opinions on when to wash herbs. Some gardeners wash them immediately before storage; others wait until the herbs are used. I am one to wash my herbs as soon as they come into the house. In my experience, washing herbs in cold water and then spinning out the moisture with a salad spinner removes dirt and debris as well as organisms that feed on plant decay. If you don't have a salad spinner, pat the herbs dry with paper towels.

Storing fresh. There are two groups of herbs: soft and woody. The soft herbs, like basil, cilantro, and parsley, have soft stems and tender leaves and typically keep for 4 to 7 days. The woody herbs, such as oregano, rosemary, and thyme, have hard or woody stems and typically keep for 1 to 2 weeks.

When fresh herbs start turning dark or brittle or you see signs of mold on the stems, it's time to toss them.

Fresh herbs can be kept on the countertop much like fresh flowers. Trim the stems and remove the bottom leaves before putting them in a container with an inch or two of water. Keep them out of direct sunlight and change the water every couple of days. Basil, mint, cilantro, dill, and parsley are perfect candidates to keep on the counter because they quickly soften and decay in the chill of the refrigerator.

To refrigerate any herb, wash it, then loosely spread out a few sprigs between sheets of paper towels and seal in a plastic bag before storing in the fridge.

Freezing. Several simple methods retain fresh herb flavor: freezing herbs in oil or water, on their own, or as an herb butter.

Freeze in oil: Basil, rosemary, sage, thyme, and oregano are some of my favorite herbs to freeze in organic extra-virgin olive oil.

Carefully remove the herbs from their stalks and roughly chop into smaller pieces. Fill the wells of an ice cube tray halfway with herbs and then top them off with oil. Freeze for at least 8 hours, then remove the cubes and store in an airtight container in the freezer.

Freeze in water: Parsley, cilantro, and chives do well frozen in water.

Prepare the herbs based on how you will use them (chives would be chopped into small pieces, for example). Fill the wells of an ice cube tray one-quarter full of herbs and top off with fresh water. Freeze for at least 8 hours, then remove the cubes and store in an airtight container in the freezer. When you are ready to use your herb cubes, place them in a strainer over a small bowl and let the water melt, or add the cubes directly to your recipe if the extra water is not an issue.

Freeze without a liquid: Line a baking sheet with parchment paper and spread herb cuttings on top. Place the sheet in the freezer for at least 8 hours, then transfer the herbs to an airtight container and store in the freezer.

Freeze as an herb butter: Soften a stick of butter. Add herbs (a popular combination is sage, thyme, and rosemary) to the butter and mix by hand or with a hand or stand mixer. Transfer the herb butter to a piece of parchment paper, roll the paper up tightly into a log, and place in a zipper-top plastic bag. Store in the freezer.

Air-drying. The best herbs for this method are bay leaves, dill, lavender, oregano, rosemary, sage, summer savory, and thyme. Dried herbs are ideal for making teas or seasoning blends like herbes de Provence. Tie freshly harvested, washed, and well-dried herbs into bundles with twine and hang them by the stems upside down in a cool, dark place. Alternatively, place a bunch of herbs in a paper bag, clip the bag closed, and place the bag in a cool, dark place.

Check the herbs after 2 weeks. If they crumble to the touch, they are officially dry enough. Transfer them to an airtight container and secure the lid.

Dehydrating. Dill, sage, mint, oregano, basil, rosemary, lavender, tarragon, lemon balm, thyme, and sage work well. I recommend using a home food dehydrator, which can be set at lower temperatures than many home ovens and uses less energy. (When temperatures are too high, the herbs' essential oils dissipate, and you lose flavor.) Load the machine's trays with herbs, allowing just enough space between leaves or stems for air to flow, and rotate the trays occasionally—sometimes different trays dry faster than others. The average dehydration time is 6 to 8 hours, but the actual time depends on how much moisture is in the plants. When the herbs are completely dry, store them in airtight jars.

Making herb-infused vinegar. Use balsamic, rice, apple cider, or white wine vinegar to preserve and use your fresh herbs in a delicious form. Infused vinegars go great over salads or as a dip for warm bread.

Begin with a mason jar with a plastic lid (vinegar will corrode metal). One method is to pack the jar halfway with herbs and pour vinegar over the herbs to fill the jar. The other approach is to use ½ cup of herbs to 2 cups of vinegar. Either way, let the herbs infuse for several weeks. The longer the infusion sits (up to 6 weeks), the more intense the herbal flavor.

Once your vinegar has the desired flavor, strain out the herbs and pour the mixture into a bottle. Using a bottle with a cork stopper is the best choice (again, you don't want the vinegar to corrode a metal lid). If you plan to store the vinegar for more than 6 months, coat the cork with beeswax to make it airtight.

Day 7

James Clear, author of the bestseller *Atomic Habits*, says the cardinal rule of behavior change is this: What gets rewarded, gets repeated.

So today is the day to treat yourself. Make the most delicious meal, invite your friends and family, and sit down to a table full of flowers and food you grew yourself. Savor each bite, each smell, the color, and the sounds of the people you love most eating the food you grew. Moments of celebration like this will keep you digging in next week.

Perhaps you start with a vibrant salad featuring peppery arugula, spicy mustards, and tender Swiss chard, tossed with a dressing infused with fresh chives and oregano. For the main course, consider a robust kale and white bean soup, simmered with a bouquet garni of rosemary, thyme, and sage—perfect for those cooler evenings.

Roasted dishes can be a highlight too; think of a mix of cherry tomatoes, zucchini, and peppers, roasted to perfection with a sprinkle of basil and a drizzle of olive oil, ideal as a side dish or a topping for a warm, crusty bruschetta.

For a comforting crowd-pleasing pasta dish, prepare a homemade tomato sauce, simmered with garlic, oregano, and a hint of rosemary, tossed with al dente pasta, and topped with freshly grated Parmesan.

Or should we make tacos? Create a colorful filling of sautéed squash, beans, and peppers, seasoned with cumin and fresh cilantro, and served in warm tortillas for a delightful and healthy twist on taco night.

Each of these dishes not only brings the freshness of the garden directly to your table but also allows the natural flavors of each vegetable and herb to shine through, creating a truly homegrown culinary experience.

gardener time

"The garden is the first order of the day before I leave for work. After 3 days away, it's the first thing I tend to before bed."

— @funnygyrl

DAYS

What's an ideal warm season day? You start with a slow morning stroll through the garden pathway, listening to the birds, watching the sunlight sparkle on the dewy leaves, and taking in deep breaths.

You can't help but touch a few leaves as you walk by. Maybe you pick some kale, chives, basil, and arugula too. And if there's a ripe tomato within reach, it's definitely headed for your mouth.

A few minutes later, you've grabbed Swiss chard, a few peppers, some tomatoes, and more herbs. And even though you're cutting food for breakfast, you're also telling your plants to keep producing.

You grab a few fresh mint leaves and rub them between your fingers to inhale their fragrance. Then you head back inside. As water boils for tea, you throw in the mint leaves, lay the morning's bounty in the sink for a quick rinse, and grab your garden journal. You take a sip of tea, write down a few notes, and finally rise to make yourself a quick garden omelet with the freshly harvested tomatoes and peppers, chopped chives, basil, and of course, a little cheese.

In just a few minutes, you've moved your body, had a time of reflection, made plans for the day, nourished your body with good fresh food, and even tended the garden.

And that's enough. Even if you don't return to the garden or prepare fresh food the rest of the day, you've already had a garden-centered experience before the sun has fully risen.

"As I drink my coffee in the early morning hours, I check my plants first thing."

— @klwilson10

Morning

Mornings in the warm season are some of the best moments. You can step outside as soon as you wake up and find something to watch, tend, cut, and bring inside.

7:00—Start with a cold cup of tea: Use herbs you've grown and dried, such as mint, rosemary, or lemon balm.

7:01—Journal your top gardening goal: Cold cup in hand, grab your journal and write down the expected high and low temps for the day and list your top goal for the garden that day based on the month and day. If the temperatures allow it, plan to plant something outdoors.

7:02—Head out to the garden: Take a quick walk around and see what's happening.

7:03—Prune a few stems: You might focus on those that are in the pathway.

7:04—Grab some peppers and greens: Bring them inside for breakfast.

7:05—Make a quick breakfast: Feature the freshly harvested kale, Swiss chard, tomatoes, peppers, and herbs.

Your morning cup of herbal iced tea isn't just a drink to start your day, it's a reminder that your 5 minutes in the garden today will be so worth your time.

While you sip, note the changes in the garden, because things are changing daily. Growth will never be this rapid again. Decide on the main activity for the day—whether it's planting, tending, or harvesting—and make sure you map out the 5 minutes you need to get it done.

If you started seeds for a second warm season planting or the upcoming cool season, now is the time to check on them. Turn on indoor grow lights, ensure the seedlings have enough water, and confirm that the airflow and temperature are good too.

Cook up a colorful omelet with fresh tomatoes, peppers, basil, and other herbs, or make a garden salad for lunch with greens, sprouts, and a variety of summer vegetables. In just a few minutes, you've had time to enjoy a fresh tea, jot down some notes, make a plan, and harvest a little something to enjoy later that day.

Create a Ladybug Home

Attracting ladybugs to the garden is one of the best ways to control pests and maintain a healthy ecosystem. Ladybugs not only make you feel like you're surrounded by best friends in the garden, they're also serious eaters of aphids and other insects that love to eat greens from your garden.

To welcome ladybugs to visit and stay in your garden, focus on providing their preferred food sources.

Ladybugs thrive in a diverse garden with a mix of plants that bloom at different times of the year for a continuous food source. This diversity also supports a wider range of insects and microorganisms, which in turn creates a more balanced and resilient garden ecosystem. In particular, ladybugs are drawn to flowers that provide easy access to pollen and nectar, like yarrow, dill, fennel, and cilantro. Ladybugs enjoy the pollen from calendula, cosmos, marigolds, alyssum, dandelions, and geraniums. Additionally, they are attracted to the scent of marigolds and chives as well as the pests those plants naturally repel.

Ladybugs look for shelter in leaf litter, under rocks, or in other natural debris. Keep a small, undisturbed area in your garden to provide the shelter they need. Ladybugs often lay eggs in areas where aphids are plentiful, as this provides a food source for their larvae. To encourage this, maintain a garden that balances pest control with leaving some aphids for ladybugs to feed on.

Above all, to welcome ladybugs in your yard, you've got to skip the pesticides. Synthetic chemicals harm both adult beetles and their larvae. Opting for organic gardening practices is crucial.

By planting the right flowers, creating a hospitable environment, and adopting organic gardening practices, you can encourage ladybugs to visit and establish themselves in your garden. Their presence not only aids in pest control but also adds to the biodiversity and health of your garden space.

Noon

Midday is the perfect time to habit stack—use your lunch break as a reminder to connect to the garden.

12:00—Hydrate: Take a moment to savor refreshing sips of herb- or fruit-infused water. This could be a simple concoction of garden herbs like mint or rosemary in a glass of water.

12:01—Step outside: Take a quick, refreshing break from any indoor activities and reconnect with nature.

12:02—Plan dinner: Think about what's growing in the garden right now and about a meal you could make tonight that makes this harvest the center of the dish.

12:03—Do a quick pruning session: If you're at home, dedicate the next 2 minutes to pruning or tending seedlings. Cut off dead leaves, remove yellowed leaves, or thin a few plants to be sure each seedling has the room it needs to flourish.

12:05—Return indoors: Conclude your quick garden break and head back indoors with a few of the prunings to display in an arrangement or to cook or chop up for lunch or dinner.

In the heat of the day, you likely won't have time to do much. But if you have a free moment, one of the simpler yet important tasks is removing spent leaves and debris. Not only does this keep the garden looking tidy, it also helps prevent the spread of diseases and pests that can thrive in decaying plant matter.

With the sun at its highest, it's easier to spot and gently remove leaves without disturbing the rest of the plant. This activity doesn't require strenuous effort, making it ideal for the warmer part of the day.

The heat of the day is also a good time to observe the condition of your plants. Look for signs of wilting or stress, particularly in younger or more sensitive plants. If any plants struggle with the heat, it's a cue to adjust the watering schedule, either by increasing the amount of water or watering more frequently, preferably during the cooler hours of the early morning or late evening.

As plants grow quickly this season, their sprawling stems and leaves will often need additional support. Carry some twine in hand and gently tie up plants to nearby trellises or supports. With just one or two knots, you've made room for better light and air circulation and prevented damage to plants that might otherwise bend or break under their own weight.

Whether you're at home or at the office, you can easily turn a simple lunch into a five-star meal. During the morning routine or your quick noontime walk, snip a handful of herbs and grab some cucumbers and peppers for a refreshing and healthy midday meal. Finely chop the herbs to create a delicious sauce for pasta, sprinkle over grilled vegetables, or use in a dressing for a garden salad. Fresh herbs not only enhance the taste but also add a burst of color and aroma.

Or pick a few vegetables for a fresh veggie platter. Bell peppers, cherry tomatoes, and cucumbers can be sliced and arranged beautifully on a plate, perhaps accompanied by a dip or hummus within a few minutes

Taking the garden-to-table approach for a midday meal is incredibly satisfying and seriously healthy. The vegetables and herbs are at their nutritional peak, having been harvested at just the right moment. If you're tired of the classic sad desk lunch, then this new way of seeing the midday will change everything.

REAL FAST FOOD

Make the most of your harvests. Here are quick meals and snacks you can make with all the real food you've grown.

- *Cucumber Mint Cooler:* Blend cucumber slices, fresh mint leaves, and water; strain; sweeten with a touch of stevia.

- *Kale and Green Apple Smoothie:* Blend kale, green apples, and celery with water for a nutritious smoothie.

- *Echinacea Immune Booster Tea:* Steep echinacea leaves in hot water for an immune-boosting tea.

- *Minty Cucumber Infused Water:* Combine cucumber and mint in water for a cool, hydrating infusion.

- *Basil Lemonade:* Mix basil, lemon juice, water, and stevia for a unique lemonade.

- *Rosemary Iced Tea:* Steep rosemary in hot water, chill, and sweeten with honey if desired.

- *Stevia-Infused Herbal Tea:* Add stevia leaves to your favorite herbal tea for natural sweetness.

- *Celery and Pineapple Smoothie:* Blend celery and frozen pineapple with water for a refreshing smoothie.

- *Tomato Juice with Basil:* Juice tomatoes, blend with basil, and add a pinch of salt for a savory drink.

"I habit stack: boil water and steep tea. Then I walk the garden to see if I need to come back and do more later."

— @lisamperk

Evening

Use your time at the kitchen sink to prompt your evening routine. As soon as you've finished the dinnertime cleanup, begin your 5-minute gardener habit.

7:00—Rinse sprouts: Run the tap, slide your sprouts under the water, and ensure all the sprouts get thoroughly rinsed. Rinse out the draining tray, put the top on, and slide the tray back to its spot.

7:01—Boil water for garden tea: Into a teapot or mug, toss some dried herbs—anything from mint to chamomile, depending on what you have available. Top with boiling water.

7:02—Step outside: Take one last stroll outdoors into the garden, brush off things that may be in your way, and make space for more seeds tomorrow; take a few cuttings to bring indoors.

7:03—Sit down: Take a moment to take a seat and enjoy the sounds of the warm season garden.

7:04—Plan in your journal: With your tea steeping, take a minute to write in your garden journal. Use this 1 minute to note what's working or not working in the garden right now.

7:05—Conclude with tea: Now strain your tea, take a deep breath, then sip. Write one line of gratitude in your journal, capturing a moment from the day that's worth treasuring.

After dinner, as the intensity of the day's heat dissipates and the light softens, take a moment to walk through your green space. This is a walk for enjoyment, yes, but you can use it for planning too. Reflect on the successes and challenges of the day and plan for the next stages of the garden's life cycle.

If there's no time to plant in the mornings, the cooler evening hours are an ideal time to add new plants or seeds to the garden. The lower temperatures and softer sunlight are less stressful than the harsh midday sun for new plantings, giving seeds a better chance to germinate and young plants a gentler time to acclimate. Additionally, the evening provides a peaceful and unhurried environment for the gardener to work in, allowing for thoughtful consideration of what to plant next.

Notice the spaces that have opened up where earlier crops have been harvested; bring along a packet or two of seeds to put into empty spots. You might choose to toss in fast-growing greens like arugula, which can be harvested within weeks, or perhaps you decide to plant a second round of crops that mature later in the season, like beans. Of course, there's always more room for flowering plants or herbs.

A relaxing evening stroll through the garden is an integral part of the gardening routine. It's a time for replenishment and renewal, both for the garden and you the gardener, as you work under the fading light to sow the seeds of tomorrow's growth. This quiet, reflective time in the garden helps deepen your connection to your patch of earth and expands your understanding of the cycles of growth and change.

Enjoy the Warm Season

It was our best summer ever. And it was all because of a simple white poster I filled up as soon as the kids got out of school.

Our Summer Bucket List poster was a combination of my ideas, my kids' wishes, and a few weird challenges I found on the Internet. We placed it above the kitchen desk and checked it off, as the long days of summer passed all too quickly.

- Watch fireflies—check
- See the sunset together as a family—check
- Pick strawberries—check
- Make homemade mint ice cream—check
- Sell tea and lemonade made with herbs from the garden—check
- Grow a watermelon—[not yet checked]

The list continued, and it was my daily reminder to not let one day pass into another without taking notice. The kids kept track too, and by the time the sun started setting a little sooner, we had checked nearly all the boxes.

It truly was one of the best, but also one of the simplest, summers. And the few boxes we didn't check (successfully grow a watermelon may or may not have been one) were all the more reason to anticipate the next summer.

The warm season isn't a "to-do list" but instead an "enjoy list." It's a reminder to relish the long days, to celebrate every little thing, and to savor the moments while they last, because flavors and feelings like this are too good to go unnoticed.

QUICK PICKS

Pick and choose from the following list when you need a quick idea or direction to make the most of any free moment in the warm season.

MONTH 1

Planting

- Bring tomatoes, peppers, and squash plants outdoors to harden off.
- Chart the weather forecast for the last day of frost.
- Deep water seedlings.
- Remove bolted or spent cool season plants, like arugula, spinach, and radishes.
- Compost healthy, disease-free plant debris from the cool season.
- Add a few inches of organic compost to the soil to prepare for warm season plants.
- Do a soil test.
- Mix soil amendments into the garden.
- Plant tomatoes and peppers the first day after the last frost.
- Plant peppers and eggplants 5 days after the last frost.
- Plant cucumbers by seed under a trellis.
- Add bush bean seeds under large plants.
- Add arugula seeds under medium-size plants.
- Monitor soil moisture and water.
- Plant basil alongside tomatoes and peppers to repel pests and enhance flavor.
- Add marigolds around the border of the garden.
- Add zinnia seeds on the outside of the garden.
- Add coreopsis and cosmos to the outer edges of the garden.

- Water newly planted herbs and flowers generously.
- Research common pests in your area and their preferred plants.
- Plant trap crops like nasturtiums around the edge of the garden.
- Start hot season plants indoors (if applicable).
- Add more bush bean and arugula seeds to open spots.
- Keep soaking sprouts indoors for a continuous supply.
- Plant microgreens trays.
- Plant additional cucumber seeds alongside the trellis.
- Plant additional bush bean seeds around the garden's open spot.
- Start cool season plants indoors (if applicable).
- Start a new set of sprouts indoors.
- Start a new tray of microgreens indoors.

MONTH 2
Tending

- Deep water warm season plants regularly.
- Avoid daily shallow watering.
- Water when top inch is dry.
- Use morning or evening sessions.
- Feed with organic amendments.
- Provide nitrogen, potassium, and phosphorus.
- Ensure 6 to 8 hours of daily sunlight.
- Prioritize morning sun in partial shade.
- Maintain balance of water, nutrients, and sunlight.
- Perform regular check-ins for changing needs.
- Support vining plants with trellises.
- Tie main stems weekly.
- Put in stakes for peppers, eggplants, and squash.

- Use afternoon shade in hot climates.
- Prune vining plants regularly.
- Remove yellow or damaged cucumber leaves.
- Prune cucumber lateral shoots.
- Avoid pruning more than a third weekly.
- Trim unhealthy stems or overcrowding.
- Remove lower tomato leaves.
- Decide on pruning tomato suckers.
- Prune tomato suckers when small.
- Trim top of tomato plant late in season.
- Use clean, sharp pruning shears.
- Prune in the morning.
- Regularly check plants for pruning.
- Avoid overpruning to prevent stress.
- Protect garden from pests and critters.
- Add herbs and flowers for beneficial insects.
- Prune affected leaves, clear soil, and add compost when pests appear.

MONTH 3
Harvest

- Harvest green herbs.
- Harvest thyme by snipping stems.
- Harvest sage.
- Snip oregano stems above leaves.
- Harvest basil when leaves are abundant.
- Cut outer leaves of mustards.
- Gently dig potatoes when foliage yellows.
- Lift one garlic bulb to test readiness.
- Harvest cucumbers when firm.

Warm Season

- Pick summer squash when young.
- Harvest cherry tomatoes when ripe.
- Pick pole beans when firm.
- Harvest eggplants when glossy.
- Look for watermelon signs of ripeness.
- Cut cantaloupe from the vine.
- Harvest large gourds when stem dries.
- Make infused vinegar.
- Make basil ice cubes.
- Cut arugula to make a salad.
- Hang herbs to dry.
- Store dried herbs in jars.
- Dehydrate peppers.
- Make sun-dried tomatoes.
- Make garden-fresh pico de gallo.
- Make refrigerator pickles.
- Make green juice with kale and cucumbers.
- Make a sun-dried tomato bread.
- Make pasta with zucchini noodles.
- Create a stir-fry from garden harvests.
- Host a warm season harvest party.

"I've noticed that on the weekdays when I do make it out to the garden, it's generally because I've squeezed it into my mornings before work and told myself I'd only be out there for 5 minutes. Sometimes I'm out there much longer, as time in the garden tends to reprioritize my morning routine, but on really tight mornings, 5 minutes will do. I like to go barefoot, and my days are so much more intentional and focused when they start in the garden."

— **Lauren Liess,**
author of Down to Earth, Feels Like Home, Habitat, *and* Beach Life

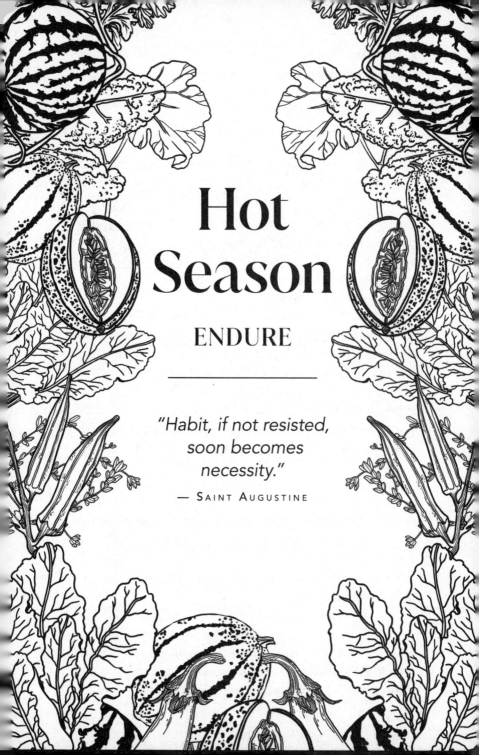

Hot Season

ENDURE

---

*"Habit, if not resisted,
soon becomes
necessity."*

— SAINT AUGUSTINE

Some might say that gardening isn't possible (or enjoyable) when daily temperatures exceed 95°F in the heat of the day and remain 80°F or higher at night. (And some might say they're right!)

And it's true: gardening in high temperatures is challenging. We may be used to eating foods that thrive in the warm season, when temperatures are milder and more predictable, but we can borrow plants and practices from tropical or subtropical regions to keep the gardening habit going through even the most intense temperatures of the year.

That's why I look to North African, Caribbean, and South Asian cuisine for inspiration because these regions show us that loads of leaves, roots, and fruit can grow during the hot season. The leaves include herbs as well as greens (like Malabar and New Zealand spinach) uniquely suited for high temperatures and unpredictable rainfall. The key root to grow in this season is the sweet potato. The fruits include hot peppers, eggplants, and tomatillos as well as special types of cucumbers, beans, and okra.

Is there enough here to keep your garden full all through the hottest part of the year? Absolutely. And you'll want to do just that—the less soil exposed to the hot sun, the better. Even if you only cover your garden with the spreading foliage of sweet potatoes, you'll have both leaves and roots to harvest and delicious meals to enjoy the minute you come inside.

If you're not gardening close to the equator or at a high altitude, you may not experience a hot season in the garden. But as the climate changes, the likelihood of experiencing months with temperatures exceeding 95°F and unpredictable rainfall is increasing—for all of us.

What's Growing Outside in the Hot Season

Leaves
Arugula
Basil
Collard greens
Kale
Malabar spinach
Mint
Mizuna
Mustard greens
New Zealand spinach
Oregano
Rosemary
Sage
Swiss chard
Thyme

Roots
Ginger
Sweet potatoes
Turmeric

Fruit
Armenian cucumbers
Eggplant
Hot peppers
Muskmelon
Okra
Tomatillos
Yard-long beans

Flowers
Angelonia
Coreopsis
Cosmos
Sunflower
Zinnea

Hot Season at a Glance

MONTH 1: Planting hot season plants and seeds, starting second season plants indoors

Week 1: Prepare the garden and start second season plants indoors.

Week 2: Move hot season plants outdoors.

Week 3: Plant roots and small fruit plants outdoors.

Week 4: Plant hot season leafy plants outdoors.

MONTH 2: Tending hot season plants
>   Week 1: Feed hot season plants.
>   Week 2: Support hot season plants.
>   Week 3: Prune hot season plants.
>   Week 4: Defend hot season plants.

MONTH 3: Harvesting hot season plants
>   Week 1: Harvest hot season leaves.
>   Week 2: Harvest hot season roots.
>   Week 3: Harvest hot season small fruit.
>   Week 4: Harvest hot season large fruit.

gardener time

*"I get up early and check and weed
my garden before my family gets up."*

— Erica

MONTHS

If you have 3 months in the hot season, you'll divide them into planting, tending, and harvesting. The first is most crucial to ensure your plants are fully settled before it's time for them to grow in. Once the right plants are established for the hot season, it's just about tending and harvesting in the second and third months.

HOT SEASON/MONTH 1: PLANTING

The key to planting for the hot season is being on time. And by "on time," I mean *early*. As temperatures soar, you're in a race to beat the heat when it comes to getting new plants settled in the garden before the mercury really begins to rise.

Noting the weather as one of your daily practices helps you know when it's time to start harvesting your warm season plants more aggressively so there's room for hot season plants to go in their places. If you can get the heat-loving plants established in the garden before the temps rise and stay in the high 90s, you're 90 percent more likely to have a thriving and full hot season garden.

Hot Season/Month 1/Week 1

At this point, your garden is likely packed with loads of warm season plants like tomatoes, peppers, cucumbers, and kale. So this first week of the hot season is the time to harvest all that's possible and make space for new plants for the hot season, especially for the large plants.

Start by focusing your harvest especially on warm season cucumbers, zucchini, squash, and other melons, then tomatoes, bell peppers, and other small fruit. Then harvest any remaining root crops like potatoes, carrots, beets, or radishes, even if they're not completely finished. These plants do not fare well when temperatures climb into the high 90s. Finally, harvest arugula, mustards, any other greens showing signs of stress or that have been in the garden for more than 90 days. Now cut these plants at the base or (in the case of nightshade plants like tomatoes) pull the entire plant, root and all.

News flash! Don't pull everything. Biennial plants like kale and Swiss chard can hang on through the hot season, especially if you stage taller plants around them to provide shade. Perennial herbs like oregano, rosemary, thyme, and sage can thrive through the hot season as well. There's no need to remove those plants unless they're showing signs that they are less than healthy and no longer productive.

You'll now have pockets of blank spaces throughout the garden. To these empty spots, add a bit of mushroom compost and earthworm castings. If your plants have not grown as well as you'd hoped they would this past season, this is a good time to take a soil sample to send off for testing (see page 86).

And it's time to gather or order seeds for your second season plants. You'll need to start planting the following seeds indoors as soon as possible to be ready for the second warm season. (Note: This only applies if you have a hot season where you garden. If not, follow the steps for a second cool season; see page 243.)

Leaves
Basil

Fruit
Bush beans
Cucumber
Eggplant
Peppers
Pole beans
Squash
Tomatoes
Zucchini

Flowers
Coreopsis
Cosmos
Marigold
Nasturtium
Zinnia

gardener
time

"In my mini breaks, I'll start a coffee, run out to deadhead something, come in and grab coffee, then head to work."

— Carrie

Give Me Five

Here are quick tasks to make the most of 5 minutes this week:

- Harvest tomatoes and cucumbers left on the vines.

- Harvest root crops like beets and carrots.

- Harvest potatoes.

- Clear one garden space.

- Compost leaves and plant materials.

- Order seeds for the next season.

- Make a big dish with your harvests.

Hot Season/Month 1/Week 2

Once you've cleared enough plants to create some empty spots in the garden, start planting. Begin with the fruiting plants—okra, eggplant, tomatillo, hot peppers, Armenian cucumbers, yard-long beans, and muskmelon. They take the longest to mature and produce, at least 75 to 90 days after transplanting. Each of these plants originated in areas with hot temperatures and developed growing habits that help them survive extreme weather.

Originally grown in Ethiopia and Egypt, okra has a deep root system that accesses moisture, even in dry, very hot conditions. Plant okra by plant or seedling at the back of a garden bed or in a row down the middle of the garden. These plants grow more than 6 feet tall and need to be in a place where they won't crowd or overwhelm the other plants.

Eggplant thrives in warm weather, and its fruit develops best when temperatures are high.

Hot peppers, from the plant genus *Capsicum*, have small, waxy leaves that reduce water loss and protect against excessive sun exposure. Hot peppers produce enzymes, such as heat-shock proteins, that protect their cells from damage caused by extreme temperatures.

Tomatillos, originally grown in Mesoamerica, use deep root systems to access water in stressful times. The compact and bushy plants conserve moisture and fit into a variety of garden spaces. The fruit's papery husk protects it from sunburn or sunscald and provides a physical barrier against pests and disease.

Plant eggplants in the row with okra or just in front, as these plants grow tall but not nearly as tall as peppers. Use tomato cages or garden stakes to support the eggplants as they grow. Similarly, tomatillos can be planted in the same row as eggplants or near trellises that are 4 feet tall. Install hot peppers in front of the eggplant and tomatillos, giving each one about ½ square foot of space for growth.

In the cucurbit (gourd) family, Armenian cucumbers can last through the heat of summer and be grown along tall trellises, especially arches, and muskmelons do their best to sprawl throughout the garden space. Muskmelon, also known as cantaloupe, originated in modern-day Iran and India and provide a sweet and juicy treat in the hot season.

Muskmelons have complex root systems that access water deep in the soil, and their leaves, which are covered in fine hairs to reduce water loss, have a high water content that keeps the plant cool. While most fruit do not pollinate in hot temperatures, muskmelon flowers are heat resistant and continue to fruit even when temperatures exceed 95°F. Plant Armenian cucumbers and muskmelons by seed directly in the garden.

Even though all these fruiting plants are well accustomed to the heat, you must plant these heat-tolerant varieties well before temperatures hit their hot season peaks. Plants that grow well in the hot season are ones that got established during the warm season, settling their roots deep into the soil before the blazing sun dries the soil and tests their limits.

After planting, be sure to water thoroughly to settle the soil around the roots and provide the hydration plants need to establish themselves. In the following days, regular watering, preferably in the early morning or late evening to minimize evaporation, is crucial for the plants to develop strong root systems capable of withstanding the extreme heat.

Give Me Five

When you've got 5 minutes, choose one of these tasks to accomplish for the week:

- Plant okra down the center of the garden.
- Plant eggplant in rows in front of the okra.
- Plant tomatillo in rows or at the edges of the garden.
- Plant Armenian cucumber seeds along the trellis.
- Plant muskmelon to grow in the pathways of the garden.
- Plant hot peppers in front of the larger plants.

*"As a mom of a three-year-old and a 10-month-old,
I get out during babies' nap times."*
— Chantelle's garden diary

Hot Season/Month 1/Week 3

The third week of the planting month is the perfect time to plant root crops for the hot season. And the winner in sizzling temperatures is the sweet potato.

In fact, even if you want to take a break from the heat of the hot season garden, you should consider planting sweet potatoes. Sweet potatoes are considered one of the oldest cultivated crops in the world, possibly dating as far back as 4000 B.C.E. and are believed to have originated in the tropical regions of Central and South America, particularly in what is now Peru and Ecuador. Sweet potatoes have extensive root systems that allow them to access moisture deep in the soil, so they're relatively drought resistant, which is crucial for surviving in hot and arid environments.

Sweet potatoes are easy to grow in a variety of soils, from sandy to loamy, they can tolerate a wide pH range, and they're relatively resistant to pests and diseases that can be more prevalent in hot and humid climates. These plants store energy in the form of starches and sugars in their tubers, which allows them to persevere during dry seasons and provide a reliable food supply in hot climates.

Sweet potatoes are typically grown from "slips" (shoots). You can buy these from a nursery, or create your own by placing a sweet potato in a container of water. Once the potato produces shoots, you can remove them and plant them in the garden.

Grow sweet potatoes alone in a raised bed (not interplanted with others) or alongside the garden area directly in the ground. These slips and starts start small, but the vines eventually take over the entire garden space. Giving the plants ample room now prevents losing other plants later.

To plant sweet potatoes, dig holes or trenches 4 to 6 inches deep and 3 feet apart. Place the slips in the holes with the top leaves above the soil surface. Space each slip 12 to 18 inches apart within the rows.

Hot Season/Month 1/Week 4

No bare soil. That's the mantra for a thriving garden in the middle of the hot season. And this happens when you plant additional crops at the base of larger plants. This intensive planting technique not only keeps the soil covered but also helps retain moisture, reduce soil temperature, and suppress weeds. Plus, you get more food from your garden!

The best plants to cover the soil in your garden include peas, beans, greens, and herbs. Peas and beans grow best in the middle of the garden beds between plants, while greens and herbs grow best along the border and outer area of the gardens.

Lima beans and black-eyed peas are heat tolerant and beneficial for the soil. As they grow, they provide shade and cover, reducing the impact of the sun's intensity on the soil. These legumes have the added advantage of fixing nitrogen in the soil, which can enhance soil fertility and also benefit neighboring plants. These beans and peas grow relatively quickly and can fill in the spaces between larger plants, forming a protective canopy over the soil.

Crowder peas have deep-rooted systems to maintain their moisture levels, heat-resistant flowers that continue to pollinate and set fruit even when it's hot, and the ability to produce a harvest in less than 75 days so they can complete their life cycle more quickly. While these peas produce food for you, they're also benefiting the soil you'll plant in next season because of their ability to fix nitrogen in the soil. This will benefit future plants and also those currently growing nearby, like okra, eggplants, and peppers. There are so many varieties of crowder peas, including black-eyed peas, cream peas, and purple hull peas, each with a unique flavor and appearance.

Several herbs are known for their heat resistance and can thrive in the garden even when temperatures exceed 95°F. Basil is possibly the most heat-resilient herb. Some basil varieties, like holy basil and Thai basil, really thrive in hot weather. And basil can even become more fragrant and flavorful when grown under intense sun. Oregano, rosemary, and thyme also are well suited for hot and sunny locations and will continue to produce new leaves even on the hottest days. In fact, most herbs that grow through the warm season will also tolerate really hot temperatures, due to their origin in the Mediterranean climate.

Beyond herbs, you can grow heat-tolerant greens throughout the hot season. These include Malabar spinach, amaranth, New Zealand spinach, collard and mustard greens, and even arugula.

Malabar spinach is not like your typical grocery store spinach. Because it originated in hot places like India, Sri Lanka, Indonesia, and the Philippines, this plant has deep roots that can access water deep underground, a vining habit that allows it to grow tall and escape the hot temperatures of the soil, and a leaf texture that is thick and fleshy and doesn't lose much water or wilt in high heat. Malabar spinach is definitely an acquired taste, but its edible leaves are perfect for soups, stews, salads, and stir-fries.

New Zealand spinach is another green that's not actually like spinach at all but can provide another green during the hot season.

Collard greens and mustard greens are leafy vegetables with origins in different parts of the world, but both are known for their ability to grow well in hot climates. Collard greens' heat and humidity tolerance is partly due to their Southern heritage. Their large, robust leaves provide shade to their roots, helping to conserve soil moisture and regulate soil temperature. Mustard greens' heat tolerance is attributed to their ability to withstand dry spells and intense sunlight. The leaves of mustard greens have a waxy surface that reduces water loss through transpiration, making them well suited to hot weather.

And then there's arugula. You've been growing arugula since the cool season, but it's actually one of the few greens that can hang in there even when temperatures are at their highest. Arugula can be started from seed in the hot season and grown under the shade of larger, established plants like okra, eggplant, and hot peppers.

On the edge of the garden, add flowers that can hang on in the hot season: zinnias, marigolds, angelonia, sunflowers, lantana, salvia, gazania, coreopsis, verbena, cosmos, gomphrena, petunia, geranium, black-eyed Susan, and Mexican sunflower.

HOT SEASON/MONTH 2: TENDING

Put on your sun hat and drink some water, because it's time to head out and take care of your heat-loving plants. The best time to do this is in the mornings and evenings, when the sun isn't as high or hot.

If you've planted your garden intensively with lots of leaves, roots, and fruit that thrive during the hot season, then your tending tasks shouldn't make you break a sweat. Plants that have adapted to thrive in hot and dry conditions know how to take care of themselves in the garden. Your job is simply to check on them daily and ensure nature's working its magic.

Hot Season/Month 2/Week 1

Spend the first week of the tending month ensuring your plants have all they need to survive the heat. This is the feeding week, so your focus is on water, nutrients, and light.

First, ensure there's a good watering system or schedule for all your plants, especially the fruiting ones. These plants will need more than the typical prescribed 1 inch of water per week to make it through the hottest days when evaporation will be at its peak.

The extreme heat and rapid evaporation mean that gardens require more frequent watering. It's crucial to water deeply and consistently, ideally during the cooler early morning hours, to ensure that the water reaches the roots where it's most needed.

Schedule at least one day to be a deep watering day but check on the plants daily to be sure an extra watering isn't necessary. You can water by hand or simply monitor your own drip or irrigation hoses. But getting water to the roots of your plants early in the day is key to days that are full of plant growth.

Avoiding stalls in the watering patterns for your plants can be just as critical as the water itself. Plants that experience stress and long periods without water can experience cell damage that keeps the plant from growing to its full potential.

Sweet potatoes need consistent moisture, especially during the early stages of growth. Water them deeply, but avoid overwatering, as they are susceptible to rot. A soaker hose or drip irrigation system can help maintain even moisture levels.

Changes in color on the leaves or slow growth can indicate that the plants aren't getting the nutrients they need. If plants are just beginning their growth, use a nitrogen-rich amendment like compost or earthworm castings. But if the changes in color occur later in their growth, adding bonemeal or blood meal can help ensure the plants get the nutrients they need to grow to the next level.

Hot Season/Month 2/Week 2

Many of the hot season plants are sprawling or tall and may need extra support in the coming weeks.

Malabar spinach, yard-long beans, and Armenian cucumbers will need a trellis or tall structure to climb along in the weeks to come. Ensure the trellis is nearby and connect any vines that are heading in different directions back on to the trellis to keep the plant supported and avoid breakage. Muskmelons will vine as well but won't need a vertical trellis if there's room for the plant to sprawl in the walkways of the garden. I typically grow muskmelon right on the edge of a raised bed and give it space to spread out and grow in and around the garden area.

Tomatillos, eggplants, and peppers are bush-like in nature, but they will topple if not supported with a tomato cage or stake and twine. Take time now to add supports near these plants so they can have full support when they grow larger and begin to fruit in the coming month.

For crowder peas and lima beans, hill some soil around the base of the plants. Hilling supports the root system of these plants and also ensures they'll dry out less quickly.

Use a little compost or simply push the existing garden soil up and around each plant's base. Hilling can be done for muskmelon, too, to ensure the plant has plenty of support around its roots as it sprawls throughout the garden.

Hot Season/Month 2/Week 3

Week 3 is for pruning to ensure that each plant has all the room it needs in the garden, but no more.

Begin with vining plants like Armenian cucumber. Prune back the side shoots so that one main stem thrives on each plant. You can also prune off leaves that look less than healthy as well as a few others that are at the base of the plant.

Less pruning is needed for yard-long beans and Malabar spinach. For Malabar, be sure to keep all the prunings, as each part of this plant is edible and the leaves are the highly nutritious part of the plant.

Eggplants, tomatillos, and peppers will benefit from light pruning around the base of each plant to ensure that there's plenty of airflow and that branches aren't crossing one another. Peppers can be pinched back early in their growth to encourage the plant to become more bushy (and more fruitful). To pinch back a pepper plant, simply cut it right above a leaf node once the plant has reached 6 to 12 inches in height. The plant will branch out at the leaf node and create two new branches in that spot instead of just one.

Next, cut off any unhealthy leaves or branches and take a little time to inspect for disease. Because these plants are bushy in nature, they won't need a lot of pruning to help with fruit production but cutting back an excess of leaves or unproductive stems can spur more fruit production.

Peas and beans may need some pruning to corral the plants and keep them from invading other parts of the garden. Even bush variety plants still may sprawl a bit through the garden. Although the shade created by the extra foliage is beneficial for the soil, the small vines may impact other plants, so prune back the unwanted growth.

Finally, It's time to prune herbs and greens. Perennial herbs like oregano and rosemary, if planted earlier in the cool or warm season, are likely exploding in growth and may need to be trimmed to make room for other greens and flowers. Cut the branches right above a leaf node in order to keep the herbs productive.

Cut basil at the leaf nodes to maintain production and keep the plant from flowering prematurely in the heat.

Finally, prune back greens like arugula or mustards. Remove lower leaves so they don't touch the soil. This helps prevent disease and pests on the tender greens.

Be aware that even healthy plants may struggle and stop producing fruit when temperatures exceed their tolerance levels, typically around 85° to 90°F. Remove these plants to make room for more heat-tolerant varieties. This not only saves water and resources but also improves the overall health and productivity of the garden.

Hot Season/Month 2/Week 4

High temperatures can create an ideal breeding ground for various pests and diseases, so this week, focus on protecting and defending the garden.

Organic pest control methods, such as introducing beneficial insects like ladybugs (see page 184) and lacewings, can be effective in managing pest populations.

These beneficial bugs feed on common garden pests such as aphids and mites, helping to keep their numbers in check.

Diseases, especially fungal infections, can proliferate in hot and humid conditions. To manage these organically, turn to copper-based fungicides or baking soda sprays, which are effective against a range of fungal diseases. Aim to maintain good air circulation around plants to reduce humidity and prevent fungal spores from taking hold.

The good news is that many hot season plants can fight off predators themselves. The compound responsible for the fiery heat in hot peppers, capsaicin, may serve as a defense mechanism against herbivores and pests and can deter animals from consuming the plant. Tomatillos as well have a degree of natural resistance to certain pests and diseases common in hot climates, making them less susceptible to infestations.

Give Me Five

Got 5 minutes? You can do the following things this week:

- Gather materials to protect your garden.
- Prune away lower leaves.
- Add compost around the base of plants.
- Make homemade garlic barrier spray.
- Use garlic spray on affected leaves.
- Harvest herbs and greens.
- Make a fresh dish from the garden.

*"I love to use [gardening] as a pause after
my workday to transition to the
home part of the day."*

— Shyla

HOT SEASON/MONTH 3: HARVESTING

Harvesting during the hot season requires careful timing and attention to detail. High temperatures can accelerate growth and maturation of these plants, affecting their flavor and texture.

So follow this month's schedule for harvesting and make the most of the hard work you've put into the garden.

Hot Season/Month 3/Week 1

By this point, you're a pro at harvesting herbs. Don't stop now!

Continue to cut from the outer and lower edges of each plant. As temperatures rise, herb plants may become more woody in nature to protect themselves from the extreme heat. But a few trims at the top of the plant will encourage new growth that will be more soft in nature.

Begin harvesting Malabar spinach, New Zealand spinach, collard and mustard greens, and arugula this week if you haven't already started. These greens are ready to harvest 60 days from planting.

To harvest most greens, cut from the base of the plant, removing the older and lower leaves first. This is especially true for arugula, greens, and New Zealand spinach. For Malabar, you'll simply cut off a few leaves at a time. These greens should be picked when they are young and tender, as mature leaves can become tough and bitter in the heat. Regular harvesting of greens encourages continuous production and can extend the harvest season.

Give Me Five

You can do one of these tasks this week even if you only have 5 minutes:

- Pick your favorite recipes to make with fresh garden herbs.

- Harvest oregano, sage, and thyme.

- Hang half of the oregano, sage, and thyme harvest to dry.

- Harvest basil.

- Freeze basil with olive oil for future use.

- Harvest greens and mustards.

- Make a quick omelet with garden greens and herbs.

Hot Season/Month 3/Week 2

This week, look to the roots and begin harvesting sweet potatoes. The first signs of readiness are when the leaves start to yellow and wither. This usually happens in late summer or early fall, depending on when you planted them. Next, gently dig around the base of the plant with a garden fork or your hands, being careful not to damage the tubers.

If sweet potatoes are ready, you'll find that the skin is firm and the tubers have reached a good size. Carefully dig them up, striving not to damage the tubers. Cure harvested sweet potatoes by storing them in a warm, humid place for about 10 days to improve flavor and storage life.

"I work from home and garden on my lunch break. Pick weeds, clip zinnias, and get some sun!"

— Krystal

Hot Season/Month 3/Week 3

This week is all about small fruit—your first harvests of crowder peas, lima beans, and hot peppers.

When their pods have reached a full and plump state, the peas are ready for harvest, usually 70 to 90 days after planting, depending on the variety and local conditions. It's best to harvest crowder peas when the weather is dry, as wet conditions can increase the risk of mold or disease on the pods.

Grab a pair of gardening gloves and a container to hold the peas. Check the lower parts of the plant first, as the peas there tend to mature earlier. Gently grasp the stem of the pea pod with one hand and the pea plant itself with the other. Apply a slight tug, and the pod should easily snap off from the plant. If it's resistant or difficult to remove, it might not be fully mature, so leave it on the plant for a few more days. As you go along, place the harvested peas in your container, being careful not to overcrowd them to avoid bruising.

Crowder peas continue to produce pods as long as the weather remains favorable, so you may need to harvest multiple times throughout the season. Check the plants regularly, and don't wait too long after the pods have reached maturity—overripe peas may become starchy.

Once you've collected your crowder peas, you can enjoy them fresh by cooking them right away or store them in the refrigerator for a few days. Alternatively, you can dry them for long-term storage.

Hot peppers can be harvested at various stages of maturity, depending on the desired heat level and use. However, in high temperatures, it's important to harvest them before they overripen and potentially turn bitter. Frequent harvesting also stimulates the plant to continue producing more peppers throughout the season.

Give Me Five

Here's a few tasks you can do this week when you've got 5 minutes:

- Select your favorite recipes to make with hot season harvests.
- Pick crowder peas and beans.
- Store beans for future use.
- Harvest long beans.
- Cook and then store beans for later use.
- Harvest hot peppers.
- Make a large pot of peas and rice.

Hot Season/Month 3/Week 4

Finally, it's time to collect big fruit!

Harvest eggplants before they grow too large and their skins lose their glossy sheen. As eggplants mature in the heat, they can become bitter and their texture can turn spongy. Harvesting them while they are medium-size and still have a shiny skin ensures that they retain their characteristic flavor and firm texture. Regular harvesting also encourages the plant to produce more fruit.

For okra, the key to a good harvest is timing. Harvest okra pods when they are 2 to 3 inches long. At this size, the pods are soft and flavorful, without the woodiness that characterizes overgrown okra. In hot weather, okra pods grow rapidly, often requiring daily harvesting. If left on the plant too long, they can become tough and fibrous.

Tomatillos grow inside a papery husk that starts off green and tightly encases the fruit. As the tomatillo matures, it fills out the husk, which then splits or cracks, revealing the fruit inside. The cracked or slightly burst husk is one of the key indicators of ripeness. Another indicator is feel: gently squeeze the tomatillo. A ripe tomatillo should feel firm and plump, with no soft spots or wrinkling. The color of the fruit itself is typically bright green, but some varieties may turn slightly yellow when fully ripe. Don't be alarmed if the tomatillos have small blemishes or marks on the skin; these are usually harmless and won't affect the taste.

Harvest tomatillos simply by grasping the ripe fruit and giving it a gentle twist or tug. It should easily detach from the plant, leaving the papery husk behind. Tomatillos that have fallen to the ground are likely ripe and can be picked up for use as well. Remember that tomatillos can produce throughout the growing season, so you may need to harvest them multiple times. Store your freshly harvested tomatillos in a cool, dry place or in the refrigerator, but be sure to remove the husks before use.

*"I've been gardening since I bought my first house
at the age of 24 in the Portland, Oregon, area almost
40 years ago. This is when my 'garden walk'
tradition began. First thing in the morning, I would
grab a cup of coffee and peruse the garden looking for
new seedlings, growth in other plants, and anything else
that looked different from the day before. Often,
I would fall to the temptation of pulling weeds, tying
things up, and before you know it I would
still be in my pajamas, dirt covered but happy."*

— Marie

WEEKS

The work stacks up as things heat up, but by breaking the
hot season into the following daily habits and goals, you
can accomplish a lot in just a few minutes.

Day 1

In the hot season, when extreme temperatures and scarce
rainfall can severely impact a garden, checking the weather
forecast at the beginning of each week can make all the difference.

Extremely high temperatures, particularly those over
95°F, can stress plants, leading to issues like wilting, sunscald, and interrupted flowering or fruiting.

When you know a heat wave is coming, set up shade
cloth to protect sensitive plants, then adjust your watering

schedule. If little to no rain is expected in the week ahead, increase watering times to ensure deep soil saturation. This is particularly crucial for newly planted seedlings and young plants that have not yet developed deep root systems.

If your area is prone to summer storms, check for the likelihood of heavy rain or thunderstorms and secure structures like trellises, provide extra support for taller plants, or move potted plants to sheltered areas.

Also be on the lookout for increased humidity levels and wind forecasts. High humidity coupled with high temperatures can promote the growth of fungal diseases. Strong winds can damage plants or dry out the soil faster.

Day 2

This is the day to add new plants like okra, eggplant, tomatillo, and peppers that thrive in high temperatures and serve a dual purpose by providing shade and some heat relief for smaller, more heat-sensitive plants.

Okra, with its tall, leafy stalks, is particularly useful for creating patches of shade in the garden. Its rapid growth and vertical stature cast a welcome shadow over lower-growing plants, such as herbs or leafy greens, which might struggle under direct, intense sunlight. The okra plants, in return, benefit from the lower temperatures and reduced stress due to the presence of companion plants.

Eggplants have large, dense foliage that can provide substantial ground cover, reducing soil temperature and moisture evaporation around neighboring plants. This can be particularly beneficial for companion plants that require cooler soil conditions to thrive.

Tomatillos offer a considerable amount of shade. Planting tomatillos in proximity to smaller plants can help protect these companions from the harsh midday sun, reducing the risk of heat stress and sunburn.

Peppers can be strategically placed in the garden to create microclimates. The foliage of pepper plants, although not as large as eggplant or okra, can still provide essential shade to neighboring plants. Additionally, the relatively compact size of pepper plants makes them ideal for filling in gaps in the garden without overcrowding.

Day 3

When walking through the garden, look for open spots where previous crops have finished their cycle or where plants have been removed due to heat stress. These areas are perfect for sowing black-eyed peas, which grow quickly, provide a valuable crop, and improve soil fertility for future plantings. Black-eyed peas enrich the soil by fixing nitrogen, which can benefit the overall health of the garden.

Arugula grows quickly, providing a fresh and spicy addition to salads in a short amount of time. You can sprinkle arugula seeds in any available spot in the garden, between rows of slower growing plants or in areas where early-season crops have been harvested.

Plant yard-long beans along open trellises so they'll add shade and structure to the garden in the days to come. During a garden walk-through, identify areas alongside fences, trellises, or other supports that are prime locations to plant yard-long beans. The beans not only utilize vertical space efficiently but also provide a steady supply of long, flavorful pods throughout the season.

Have these fast-growing seeds with you on this day's garden walk-through to keep the garden productive and healthy.

Each plant offers unique benefits, from the soil-enriching properties of black-eyed peas and the rapid growth and distinctive flavor of arugula to the prolific yield of yard-long beans. This approach of interplanting and succession planting keeps the garden lush and productive, making the most of the available space and the challenging conditions of the hot season.

Day 4

Heavy watering and feeding of plants each week on this day allows for a consistent routine, which is beneficial for both you and the plants. Early morning is the ideal time for this task, as the cooler temperatures and lower wind speeds reduce water evaporation and allow moisture to seep deep into the soil, reaching the roots where it's most needed.

Unlike brief daily waterings that only moisten the surface, deep watering encourages the development of a strong and extensive root system, which is vital for plants to withstand the heat and dry conditions of the hot season. Water each plant slowly and thoroughly, ensuring that the water penetrates several inches into the soil. The goal is to moisten the entire root zone; for most plants, this means allowing the water to reach at least 6 to 8 inches deep.

Watering day is also the perfect opportunity to feed your plants. During the hot season, plants are in their active growth phase and require more nutrients to sustain this growth. Adding a balanced, slow-release fertilizer to the soil can provide the plants with the necessary nutrients. Alternatively, use a water-soluble fertilizer during watering to nourish the plants by delivering nutrients directly to the root zone.

It's important to observe how different plants respond to this watering schedule, as their needs can vary based on size, type, and location. Some plants may require additional watering during the week, especially if the weather is exceptionally hot or windy.

Fruiting plants, in particular, often need an extra boost of nutrients during the hot season to support their growth and fruit production. A natural soil amendment high in phosphorus like bone meal or compost can encourage more robust fruit development.

For root crops such as sweet potatoes, a dose of a potassium-rich amendment like wood ash, kelp meal, or compost can be beneficial. Potassium aids in developing strong root systems and improves overall plant vigor. It also helps in water retention, which is definitely necessary this season.

Apply soil amendments at the base of the plants, away from the leaves, to prevent burns and ensure that the nutrients are placed where they are most needed.

Day 5

The hot season can be particularly taxing on plants, making regular pruning essential.

Start by removing any leaves and stems that are no longer productive. This includes leaves that have turned yellow or brown and stems that show no signs of new growth. Such parts drain energy from the plant, which could be better utilized for new growth and fruit production.

Next, focus on removing leaves that may take energy away from the plant without contributing to its growth. This is particularly important for fruiting plants, where energy should be directed toward developing fruits rather than maintaining excessive foliage. However, it's crucial to maintain a balance, as leaves are necessary for photosynthesis and overall plant health.

While pruning, inspect the plants for signs of pests or diseases. Carefully remove and dispose of leaves that show spots, deformities, or other signs of damage. Make sure to discard these away from the garden to prevent the spread of pests and diseases. This step is critical—pests and diseases can proliferate rapidly in the hot season, potentially causing significant damage to the garden.

Spending a day focused on pruning during the hot season is crucial for the health and productivity of a garden. Pruning helps in removing unproductive growth and controlling pests and diseases. This careful maintenance ensures that the garden remains a vibrant and fruitful space throughout the hot season.

Day 6

Dedicating one day per week to harvesting from the hot season garden is the best way to make the most of all the work you've put into this space.

Focus on this task in the morning, when temperatures are cooler, so plants are less stressed and the fruits and vegetables are at their peak in terms of freshness and taste.

Start with okra. Okra pods grow rapidly and are best when harvested while they are small and tender, usually no longer than 2 to 3 inches. If left on the plant for too long, they can become tough and fibrous. Harvesting okra in the morning ensures that the pods are at their best quality, both in texture and flavor.

Next, move on to tomatillos, eggplants, and hot peppers.

Tomatillos should be picked when their husks have just begun to split, indicating they are ripe. Similarly, eggplants should be harvested when their skin is glossy and the fruit is firm. For hot peppers, the timing of the harvest depends on the desired level of heat and use—whether they are to be used green or allowed to ripen to their final color. Harvesting these vegetables in the morning ensures they retain their moisture and flavor, which can be lost in the heat of the day.

For leafy greens like Malabar and New Zealand spinach, morning harvesting is equally important. Pick these greens when they are young and tender, before the heat of the day causes them to wilt or lose their vibrancy. Early harvesting allows for the crispest and most flavorful greens.

Okra and greens, being delicate, should be used or preserved as soon as possible. They can be stored in the refrigerator if not used immediately. Tomatillos, eggplants, and peppers should be cleaned gently to remove any dirt and then stored at room temperature.

Use hot peppers for drying, pickling, or making hot sauces. Drying peppers in the sun or a dehydrator preserves them for long-term storage and intensifies their flavor. Pickled peppers are excellent for adding a spicy kick to dishes throughout the year. Homemade hot sauce, made on the harvest day, can be a rewarding way to utilize the fresh flavor of the peppers.

Make sauces or salsas with all those tomatillos. Freshly harvested tomatillos can be roasted or boiled to make salsa verde, which can be canned for future use or used fresh within the week. Their tangy flavor also makes them a great addition to soups and stews, providing a unique taste profile.

Eggplants can be roasted and stored in olive oil, made into dishes like baba ghanoush, or even grilled and frozen for later use. Eggplants freeze well, retaining much of their texture and flavor.

Malabar and New Zealand spinach can be washed and stored in the refrigerator for immediate use. These greens are excellent in salads, stir-fries, or simply sautéed with garlic and olive oil. For longer storage, blanching and freezing these greens is a viable option, preserving them without losing much of their nutritional value or taste.

Day 7

It's celebration day! Time to enjoy all that's growing in the garden. Make some delicious salads, bake a sweet potato, roast some okra and tomatillos. Have a hot season feast with roasted okra and eggplant served over Malabar spinach and arugula salad and topped with a tomatillo and hot pepper sauce. Whatever you do, don't miss the magic of taking a bite of your favorite food right there in the middle of the garden.

REAL FAST FOOD

Here are some quick recipes to make when the weather gets hot and your harvest basket is full.

- **Armenian Cucumber and Yogurt Salad:** Chopped cucumber combined with yogurt and tossed with dill, mint, and lemon juice.

- **Spicy Lima Bean and Hot Pepper Stir-Fry:** Lima beans and sliced hot peppers stir-fried, seasoned with garlic, oregano, and soy sauce, and served over steamed rice.

- **Okra and Tomato Gumbo:** Okra, tomatoes, and crowder peas cooked and flavored with thyme, oregano, and a hint of hot pepper, and served over rice.

- **Herb-Roasted Vegetables with Crowder Peas:** A mix of roasted vegetables, like eggplant, bell peppers, and tomatillos, tossed with crowder peas and a medley of rosemary, sage, and thyme.

- **Okra Fries:** Sliced okra, tossed with oil and seasoned to taste, roasted until brown and crispy.

- **Grilled Eggplant with Basil Pesto:** Slices of grilled eggplant topped with a homemade pesto made from fresh basil, garlic, and pine nuts.

- **Grilled Okra and Eggplant Skewers:** Pieces of okra and eggplant alternated on skewers, brushed with a mixture of olive oil, basil, and minced garlic, and grilled.

- **Tomatillo Salsa Verde:** Tomatillos blended with roasted hot peppers, garlic, and cilantro.

DAYS

Here's an "ideal day" in the hot season: You wake up, put your feet on the floor, and slide on your sandals. You're headed outside. You can feel the warm air on your skin the minute you open the door, and you don't even step into the garden without first grabbing the hose.

You tug at the hose and slowly move through the garden, careful to put the water at the roots of each plant. You've seen lots of scorched yards and landscapes around the neighborhood this month, but your garden is green with peas and beans, cucumbers, eggplants, and loads of peppers.

You pick a handful of beans, a few cucumbers and peppers, an armful of herbs, and a few ripe tomatillos that dropped to the ground overnight. Oh, and you're sure to grab some okra too, since you know it will double in size in the garden today.

You turn off the hose and head indoors, laying out the harvest on the counter.

You place the picked beans to soak in a bowl of water all day and make yourself a glass of iced mint tea while you cook some eggs and top them with a sliced tomatillo. You eat the most delicious breakfast and set off for the day.

You make it through work and your busy schedule until just before dinnertime, when you head back into the kitchen to grab those beans and cucumbers to make a cool summer salad. You chop cucumbers and peppers while you boil the beans. This fast dinner is finished in just a few minutes but it tastes like it took hours—because the flavors are just so perfect and fresh.

You finish the day just as the sun sets with one more step into the garden. The temperature has dropped only a few degrees, but that's just enough to make a few minutes of pruning tolerable. You cut a few leaves, stop to smell the herbs right next to your door, and head back inside.

Many people would say this season is intolerable, but you've found a way to make the most of these extremes.

Morning

The mornings are the moments you don't want to miss in the hot season—the coolest time of day to step outside and connect with the season and all that's growing.

7:00—Make herbal tea: Use mint, rosemary, or lemon balm leaves.

7:01—Journal your top gardening goal: Cold cup in hand, grab your journal and write down the expected high and low temps for the day and list today's top goal for the garden based on the month and day. If the temperatures allow it, plan to plant something outdoors.

7:02—Head outdoors: Enjoy the coolest part of the day.

7:03—Grab some herbs and flowers.

7:04—Take a picture: Capture the garden at daybreak.

7:05—Make a quick breakfast: Maybe yours features Malabar spinach, hot peppers, basil, and eggs.

Oh, thank goodness, I thought. My eyes had just opened, the sun wasn't up yet, and I could hear the soft pitter-patter on the roof. It was raining.

It was the middle of summer, temps were well over 90°F, and we'd gone the last few days without an inch of rain. The kids were busy, I was busy, and when I finally thought to water the garden, it was time to cook dinner or drive to practice, or I just needed some sleep.

I'd made a mental note to wake up and water the garden first thing the next day. But instead, I'd woken up to find out it was already taken care of. Thank you, Mother Nature!

Watering is a key task of the early morning, and your minutes of watering now are worth hours of watering later.

In the early morning, plants can absorb water quickly and store it for use throughout the key hours of sunlight. In contrast, water served to the plant midday isn't used as efficiently. As the day wears on, the plant's priorities and energy focus change, so more water mid- or late day is less likely to be absorbed and utilized. In fact, in some cases, plants have a difficult time accessing the water that's made available during an extremely hot or dry period of the day.

Carry a small bag of seeds with you as you walk through the garden. Look for open spots or gaps where these seeds can be planted. This practice helps maintain a continuous cycle of growth and ensures that your garden remains lush and productive. It also allows you to experiment with different plant varieties, making each stroll through the garden a new adventure.

In the warmth of the hottest season, iced tea made from garden herbs is a perfect start. Consider herbs like peppermint, lemon verbena, or hibiscus for a refreshing and cooling beverage. This not only hydrates you but also allows you to savor the flavors from your garden.

Spend a few minutes writing in your garden journal, recording the progress and challenges, such as which plants are coping with the heat. Make a note of any new blooms, fruiting, or signs of pests or water stress. Journaling can help you keep track of tasks and plan future garden activities.

Given the high midday and afternoon temperatures, aim to complete your main gardening activities early in the morning. This could include watering, mulching, or creating shade for heat-sensitive plants. It's also a good time to check on plants for signs of heat stress and take appropriate actions.

If you're starting indoor seedlings for the upcoming cooler season, ensure that they're doing well. Turn on the grow lights, check the moisture level of the soil, and make sure they're not exposed to excessive heat, which can be detrimental.

Don't forget to walk through the garden to pick the most ripe fruits and vegetables! Morning's cooler temperatures make it more comfortable for you and less stressful for the plants. Harvesting ripe produce regularly encourages the plants to produce more and helps prevent pests and diseases that can be attracted to overripe or rotting fruits.

Prepare a breakfast featuring produce from your garden. A fresh salsa made from tomatillos, basil, hot peppers, and other herbs can be a zesty addition to your morning meal. Consider using sprouts in a breakfast wrap or as a topping on avocado toast for added nutrition and flavor.

gardener time

"I involve my kids! My four-year-old loved helping me plant seeds in every corner of the garden— he can now name every plant that is growing and loves giving tours of the garden—and my two- year-old loves to take charge of the hose and water everything every day. This is my first year gardening, and I'm hooked on this new hobby."

— Katie

Noon

Midday is the perfect time to habit stack—use your lunch break as a reminder to connect to the garden.

12:00—Hydrate: Take a moment to savor refreshing sips of herb- or fruit-infused water. This could be a simple concoction of garden herbs, like mint or rosemary, in a glass of water.

12:01—Step outside: Take a quick, refreshing break from any indoor activities and reconnect with nature.

12:02—Grab a handful of harvest: If you're home, grab a few beans or peas, and some okra too.

12:03—Quick pruning session: If you're at home, dedicate the next 2 minutes to pruning or tending seedlings. Cut off dead leaves, remove yellowed leaves, or thin a few plants to be sure each seedling has the room it needs to flourish.

12:05—Return indoors: Conclude your brief time outside and head back indoors with a few of the prunings to display in an arrangement or to cook or chop up for lunch or dinner. If you're away from home, enjoy a prepacked lunch with herbs and snacks from the garden.

Make a cool lunch from the hot garden to cool yourself down and experience the magic of all that the sun and heat can grow in such a short period of time.

Use just-harvested fruit from the hot season, like eggplants, cucumber, okra, and hot peppers, as well as basil and greens, to craft cool salads and refreshing meals that keep you going but also encourage you to keep planting.

Grilled eggplants is a simple option. Slice the eggplants thinly, brush them with olive oil, and grill them until lightly charred. These can be served cold, drizzled with a yogurt and garlic sauce, and garnished with fresh herbs, like basil or mint, from the garden. This dish is not only refreshing but also packed with flavors that are heightened by the grilling process.

Make a chilled okra salad by lightly sautéing okra with spices, letting it cool, and tossing it with cherry tomatoes, cucumbers, and a lemon-herb dressing. This dish is crunchy, tangy, and full of textures.

Incorporating hot peppers and basil into your dishes can add a flavorful kick. For instance, a spicy basil pesto made from garden-fresh basil and hot peppers can be a great accompaniment to pasta or a spread for sandwiches.

Sweet potatoes can be roasted, cooled, and added to salads, offering a sweet balance to the spicier elements of a dish.

Finish the meal with a drink made from the garden. Melon juice, for example, is a quintessential summer refresher. Simply blend ripe melons with a touch of lime juice and mint for a cooling beverage. Alternatively, herb-infused waters are a subtle yet effective way to keep cool. You can steep herbs like basil, mint, or lemon balm in water, along with slices of cucumber or citrus fruits, for a gentle, aromatic thirst quencher.

Evening

Use your time at the kitchen sink to prompt your evening routine. As soon as you've finished the dinnertime cleanup, begin your 5-minute gardener habit.

7:00—Rinse sprouts: Run the tap, slide your sprouts under the water, and ensure all the sprouts get thoroughly rinsed.

Rinse out the draining tray, put the top on, and slide the tray back to its spot.

7:01—Boil water for garden tea: Into a teapot or mug, toss some dried herbs—anything from mint to lemon balm to anise hyssop.

7:02—Check on your seedlings and turn off the grow light: Observe their growth, check the soil moisture, and ensure the seedlings are doing well. Then turn off their grow light to signal the end of their "daylight" hours. This mimics natural light cycles, which is important for plant growth.

7:03—Plan in your journal: With your tea steeping, take a minute to write in your garden journal. Use this 1 minute to note what's working or not working in the garden right now.

7:04—Step outside: Take one last stroll outdoors into the garden, brush off things that may be in your way and to make space for more seeds tomorrow, and take a few cuttings you can bring indoors.

7:05—Conclude with tea: Now strain your tea, take a deep breath, then sip. Write one line of gratitude in your journal, capturing a moment from the day that's worth treasuring.

As the day cools down, take a leisurely walk through your garden. This might be the one moment that the garden isn't sweltering.

Evening is a great time to check the soil's moisture levels. If the garden is drying out, a second round of watering might be necessary to help plants recover from the day's heat and ensure they have enough moisture to stay healthy and keep growing.

While wandering through your garden, gather herbs for upcoming meals. Fresh herbs like basil, thyme, and rosemary can make any meal better. Use them to create vibrant sauces, or dry and store them as spices, adding a homegrown touch to your meals.

Spending 5 minutes like this gives you a deeper connection with your garden but also keeps gardening a relaxing and enjoyable ritual. It's a time to slow down, appreciate the fruits of your labor, and plan for future growth.

Grow More Basil in the Hot Season

Basil truly likes it hot. If you've only ever tasted grocery store basil, the hot season is a perfect time to try other varieties, each with its unique flavor profile and culinary use. These 10 different kinds of basil can be grown during the hot season.

Sweet basil. Perhaps the most common variety, sweet basil features large, green leaves and a classic, peppery basil flavor. It's widely used in Italian cuisine, especially in pesto and tomato-based dishes.

Genovese basil. Similar to sweet basil but with a more pronounced flavor, Genovese basil is ideal for pesto due to its slightly sweeter taste and larger, more tender leaves.

Thai basil. Known for its spicy, licorice-like flavor, Thai basil has smaller leaves and purplish stems. It's a staple in many Asian dishes, holding up well to cooking and adding a distinctive taste.

Lemon basil. As the name suggests, this variety has a citrus aroma and flavor. Lemon basil pairs wonderfully with fish and poultry and is great in dressings or infused waters.

Holy basil (tulsi). With a clove-like, peppery taste, holy basil is important in Indian cuisine and Ayurvedic medicine. It's less sweet and has a more robust flavor profile compared to sweet basil.

Purple basil. This variety is known for its striking dark purple leaves and a slightly milder flavor than green basil. It's great for adding color to salads, garnishes, and vinegar infusions.

Cinnamon basil. This type has a unique, warm aroma reminiscent of cinnamon, thanks to its cinnamate content. It's excellent for adding a twist to desserts, fruit salads, or beverages.

Greek basil. Known for its compact growth and small leaves, Greek basil has a strong aroma and is often used in Mediterranean cooking. Its miniature size makes it a smart choice for container gardening.

Lettuce-leaf basil. Featuring large, crinkled leaves that resemble lettuce, this basil variety has a milder flavor and is ideal for wraps, salads, or as a leafy bed for various dishes.

Spicy globe basil. A bushy, compact variety with small leaves, spicy globe basil has a stronger flavor than sweet basil. It's excellent for container gardening and as an ornamental plant.

ENDURE IN THE HOT SEASON

Even though I have been tempted to close up the garden in the hottest parts of the year, one look at other cultures thriving in hot weather with little rainfall taught me a key lesson—plants are dying to live, and they find ways to survive, even in the most challenging circumstances.

With deep roots, bushy structures, unique leaf patterns, and tangy flavors, the plants of the hot season will surprise you with their ability to not just survive the most extreme challenges in the garden but to thrive despite them. Just because you may not want to be out in the middle of a 100°F day doesn't mean there aren't plants that will show off and grow strong while you wait indoors for temperatures to drop.

Living a garden-centered life is about putting the right seeds in the right place at the right time, and letting the garden work its magic. With 5-minute inputs, you get to enjoy the simple everyday practices of planting, tending, and harvesting a little at a time until you find yourself lost in the middle of a beautiful, colorful, and productive garden, even when it's hot.

QUICK PICKS

Pick and choose from the following list when you need a quick idea or direction to make the most of any free moment in the hot season.

MONTH 1
Planting

- Harvest warm season tomatoes.
- Harvest warm season cucumbers and squash.
- Harvest warm season beans.
- Harvest warm season greens.
- Pull spent tomato plants.
- Pull cucumber, squash, and zucchini plants.
- Pull spent greens from the warm season.
- Add compost in empty spaces.
- Add sprinkles of earthworm castings in blank spaces.
- Rake soil to prepare it for planting.
- Make plant tags for upcoming planting.
- Plant okra in the large planting holes.
- Plant eggplant in front of okra.
- Add Armenian cucumber seeds along tall trellises.
- Plant tomatillos near eggplants.

- Plant hot peppers in front of tomatillos and eggplant.
- Add crowder pea seeds in blank spaces of the garden.
- Plant muskmelon seeds along the edges of the garden bed.
- Add yard-long bean seeds along tall trellises.
- Plant Malabar spinach along tall trellises.
- Plant New Zealand spinach seeds along the border of the garden.
- Add arugula seeds in empty spots.
- Plant extra crowder pea seeds in open areas of the garden.
- Plant angelonia flowers in the open border spots of the garden.
- Plant zinnia seeds in the open border areas of the garden.
- Plant coreopsis seeds around the edge of the garden space.
- Plant echinacea near the garden area.
- Plant rudbeckia plants near the garden area.
- Plant salvia plants near the garden area.
- Start seeds for the second season indoors.

MONTH 2
Tending

- Check water levels in the garden.
- Provide a deep water session in one area.
- Water around the fruiting plants.
- Add earthworm castings to open soil areas.
- Remove debris from under plants.
- Prune yellowing leaves from fruiting plants.
- Prune outside leaves from herbs.
- Prune extra vines from Armenian cucumbers and yard-long beans.

- Prune old leaves from an okra plant.
- Hill soil around muskmelon plants and Armenian cucumbers.
- Hill soil around crowder pea plants.
- Hill compost around the base of each sweet potato plant.
- Prune the lower leaves of tomatillo plants.
- Prune lower leaves around eggplants.
- Pinch the top of pepper plants.
- Check for pests on the underside of leaves.
- Use garlic spray on leafy green leaves marked with holes.
- Spray insecticidal soap on fruiting plant leaves.
- Check on indoor seedlings for second season.
- Dig up one sweet potato to check its size.
- Remove any spent plants from the garden.
- Attach Armenian cucumber to the trellis.
- Attach yard-long bean to the trellis.
- Tie up eggplant and tomatillo to stakes or small trellises.
- Direct trailing and vining muskmelon plants into pathways.
- Prune away bruised or marked leaves from plants.
- Pinch back zinnias and angelonia flowers.
- Cut back wilted or dried herbs.
- Pick seed heads for dried flowers.
- Rake debris from garden beds.

MONTH 3
Harvesting

- Harvest basil leaves.
- Make an herb-infused water with lemon balm.
- Make rosemary oil.

- Sauté Malabar spinach, green onions, and peppers.
- Make basil pesto.
- Harvest small eggplants.
- Cut a basket full of herbs to dry for winter.
- Pick peppers to hang dry.
- Harvest small okra fruit for a grilled salad.
- Cut okra flowers to make a kitchen table centerpiece.
- Harvest arugula for a cold summer salad.
- Make arugula pesto.
- Harvest hot peppers to make vinegar.
- Harvest tomatillos for green salsa.
- Dig up the first set of sweet potatoes.
- Pick flowers for a centerpiece.
- Harvest cucumbers to make summer pickles.
- Cut yard-long beans for a dinner stir-fry.
- Grill eggplant for lunch.
- Make okra gumbo.
- Cut mint for tea.
- Cut mustard or collard greens to make Southern-style greens.
- Cut hot peppers to make hot sauce.
- Cut basil varieties to make a basil arrangement.
- Cut flowers for pressing and arranging.
- Dig up the remaining sweet potatoes and store.
- Mix up a green smoothie with garden greens.
- Make a hot season ratatouille.
- Make herb-topped focaccia.
- Host a garden harvest party.

"Time pauses and worries fade in the garden. From the feeling of the warm sun and crunch of the pea gravel, to hearing my kids' excitement over the growth of each tiny seed, our garden has become a sanctuary for our entire family."

— Shea McGee,
founder of Studio McGee

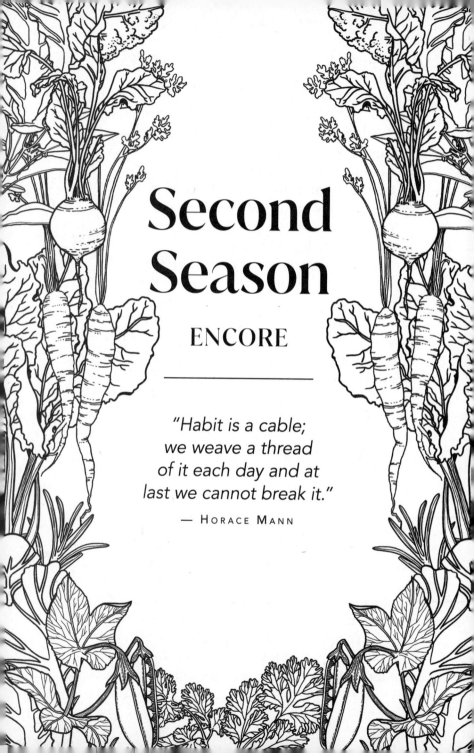

Second Season

ENCORE

---

*"Habit is a cable;
we weave a thread
of it each day and at
last we cannot break it."*

— HORACE MANN

I completely missed the tomato season my first year of gardening in Houston.

Everything I'd ever read said to plant tomatoes in the summer, so I'd done just that—headed right to the plant store, purchased tomato plants, and popped them into the garden just as school let out and the swimming pools opened.

But the brown stems and the wilting flowers were signs that my timing was off, way off.

I figured I'd done everything right—and I had, if we were living somewhere else.

It took a little research for me to learn that I was late to the Houston tomato party, more than 3 months late, to be exact. Because tomatoes were a plant for the warm season, and the sweat on my brow and the brown on those plants were both signs that *warm* was only a word used to describe the night, because the days were nothing but *hot*.

The realization was news to me and sad for my dying tomato plants. But there was some good to be uncovered.

Though June was too late for tomatoes, there was actually a second chance at tomatoes when the weather would start to turn from hot back to warm (and everyone would breathe a sigh of relief).

Turns out, Houston didn't have just one season for tomatoes. It had two. One in the early spring and another in the early fall. So even though I'd missed my chance at the start of the year, I'd get one more chance at the end.

And believe me, I took it.

This was my first introduction to the concept of the second season in the garden: the fact that each climate gets a second chance at one season in the garden each year.

For hot climates, the second season is the warm season and for cold climates, the second season is the cool one.

And for those in between, there may be a little of both—a shortened warm season and a longer cool season.

The second season is the encore—a chance to go one more round in the garden, to learn from the mirroring season's successes and lessons, and to make the most of the turns on the calendar page.

To apply the second season principle to your own garden, simply go back to the Seasons chapter of this book and find which season is next for your climate. Then look back at the section for that season and plan your days all over again.

For hot climates, this means you'll start warm season seedlings indoors during the hottest part of the year, switch out your plants to warm season as the temperature starts to drop a bit, and then work to be sure you get plenty of harvests before your weather reaches its coolest temperatures.

For cold climates, you'll start cool season plants indoors when you're in the warm season and slowly move them out to the garden just as you did previously, only this time you're replacing warm season plants instead of planting in a blank garden.

You'll do the same in mild climates, with the difference that you may not have to rush the plants as much because your cold weather won't be coming as quickly.

Wherever you live and whenever you read this book, I hope the thought of a second season encourages you. The reason the 5-Minute Gardener system works is because you actually get more chances for gardening than you'd think. Even if you miss an entire season one year because you were just too busy, there's a good chance that same season will roll around a second time.

To make the most of your 5 minutes in the garden,
head to **fiveminutegardenerbook.com**,
where you'll find Gardener Habit Trackers,
journal prompts, daily tasks, and seasonal recipes.

Conclusion

A Garden-Centered Life

"Depending on what they are, our habits will either make us or break us."

— SEAN COVEY

"So you basically don't go to the grocery store, right?"

My friend from college had seen my new business and my garden online and was convinced I was now a full-blooded homesteader.

"Not if you don't count Saturdays and Mondays," I replied.

With four growing kids at home, a business to run, and a calendar full of activities, I've never even attempted to say the words *homesteader, off the grid, homegrown,* or even *home gardener.*

I don't garden to go off grid, grow all my own food, or to avoid the grocery. I don't live to garden. I garden to live better.

Centuries ago, this book would've made everyone laugh out loud. Who would need a guide on how to fit the garden into their everyday lives? Who would need a primer on understanding their seasons or knowing what plants grow in each one? Who would need recipes and new ideas on how to make use of all the unique things that only grow in a backyard garden?

No one needed this book back then, because everyone was already living it.

The garden used to be a simple part of everyone's every day. A spot you just pass by on your way out the door, a well-worn path you walk each night before dinner, a gathering place for you and your favorite people on the weekends.

The garden used to be the source of inspiration for everything edible from drinks, sauces, and dips to full-blown end-of-the-season celebrations. For the people who have gone before us, the plants were their guide to the calendar, not the other way around.

But this is not centuries ago. We are here now, in this place where most of us, myself included, never grew up pulling a carrot or cutting some spinach before dinner. We're in a place where our food comes to us on trucks, sealed in plastic, and handed across counters.

So we now have to do a little work, we have to study a little bit, we have to go back a little to find something precious that we lost.

We have to find a way to make our time here better.

I know you're busy and stressed, and you've got a million things to do.

But can I ask you, for a little bit, to pretend that you're living a few centuries back. For at least a few minutes a day, you've got to play like you *need a garden*. You've got to act like your only source of food for the day ahead is food that you grow yourself.

You and I both know that's not true.

But what is true is this: You really do need a garden— and not for lunch, or tea, or to save on groceries. But to save you from the hustle, the rush, and the disconnectedness you feel every time you run through the grocery, honk your way through the drive-thru, or casually grab a bite of something that came from who knows where.

You and I? We may not need the garden in the sense that we can get all the food we need from the grocery. But the fact is that we've never needed the garden more in order to feel good, move more, eat better, and also reconnect to the way this whole thing works.

We both know that even though you've made it through this book, there will be days you'll skip it. There will be times you won't find the time, when 5 minutes will feel like 1 minute too much to give to the garden.

But if you try it, if you give it an actual go and give just 5 minutes of your time to the garden today, I bet you'll want to come back and do it again tomorrow.

And when you do, the food will eventually come, but the feeling is what you'll come back for. Because we were made for this, we were made through this. We as humans became who we are now because we centered our lives around gardens.

The minute you give yourself a minute to experience what it feels like to live as a gardener, you'll never want to go back to the way you were before.

So the answer to my friend is "No." "No, I don't get to skip the grocery each week." "No, I don't see myself living off the grid or making all my own food."

But yes, I do skip my way into the garden each day. I do see myself living in the middle of plants and picking a little something to awaken my senses, connect me with loved ones, and help me do real good in this world that I want to last and bloom for many, many centuries to come.

Will you join me? It'll only take 5 minutes.

Index

E

earthworm castings, 30, 38, 53, 56, 94, 112, 172, 202, 211

eggplants, 152–153, 166, 177, 204, 211, 212, 219, 221, 222, 225–226

F

fans, 29, 38, 55

5-minute gardening, 1–13, 247–249. *See also* daily habits (by time of day); month-to-month habits; seasons; weekly habits (day by day)

   flowers, 9, 29, 88, 114–115, 149–150, 159

   Fogg, BJ, 8, 58

   freezing of herbs, 178–179

   frost protection, 45–48, 51–52

   fungus. *See* pest and disease management

G

Gardenary way, 88

Gardener Time quotes and tips, 6, 12

   cold season, 26, 33, 48, 60, 66, 67

   cool season, 82, 84, 86, 92, 100, 106, 108, 117, 118, 122, 126

   hot season, 200, 202, 206, 215, 217, 220, 231

   warm season, 138, 142, 145, 146, 148, 153, 158, 161, 168, 181, 183, 189

garden fleece, 45–46

garden mesh, 98–99

garlic spray, 98–99, 112, 160

Give Me Five lists, 12

   cold season, 29, 39, 41, 56

   cool season, 87, 89, 91, 94, 96, 97, 99, 101, 103, 105, 107

   hot season, 203, 205, 214, 216, 218

   warm season, 147, 149, 152, 156, 160, 164, 167

goals, setting, 6–9, 77, 123–124

gourds, 168

greenhouses, mini, 47–48

greens, 9. *See also individual types of greens*

   cold season, 44

   cool season, 101–102, 120

O

P

Q

R

radishes, 104–105

raised beds, 46, 48, 50, 84, 207, 211

Real Fast Food, 13, 61–62, 65, 69, 188, 227

rosemary, 161

rutabagas, 105

S

sage, 162

salad, 103–104, 113, 123–124. *See also* greens

sanitizing of equipment, 55

seasons, 15–21. *See also* cold season; cool season; hot season; warm season

  second season, 18, 202, 243–245

  temperature, 85–86, 109

  year at a glance in different climates, 20–21

second planting (warm season), 170–171

seedlings, thinning, 39–40, 96–97

seeds, bulbs, tubers

  choosing/ordering, 28–29, 54, 84–85, 144–145, 202

  growing time, 81

  seed stratification, 34–35

  starting, 30–31, 37, 91

  winter sowing, 34–36, 51–52

shade cloth, 220

soap spray, 98–99, 160

soil. *See also* nutrients

  amending, 52, 146–147

  avoiding bare soil, 207

  indoor, 32

  temperature, 85–86

  tests, 86, 94, 146–147, 202

spring mix, 110, 113, 124

sprouts, 24–25, 28, 30, 53, 57

squash, 152–153, 166, 176

staggered planting times/sizes, 110, 148, 170–171

T

Index

V

vegetables. *See individual types of vegetables*

vinegar, herb-infused, 180

vining plants, 173–174

volunteer plants, 83

W

warm season, 17, 135–196

   daily habits, 182–190

   at a glance/overview, 136–139, 140, 191

   herbs, harvesting/storing/preserving, 177–180

   by month, 143–168, 192–195

   transitioning from cool season to, 140–142

   weekly habits, 168–181

water and watering, 31–32, 37–38, 41, 92, 109, 205, 210, 223, 230

weather forecasts, 169. *See also* seasons

weekly habits (day by day), 11

   cold season, 48–59

   cool season, 108–117

   hot season, 220–227

   warm season, 168–181

wind, 109

winter sowing, 34–36

Acknowledgments

Every book is a labor of so many, and there are countless of you to thank.

First, a thank-you to the team at Hay House: Patty, Lisa, Monica, and my editor, Paula: I couldn't have done this without you.

To my incredible team at Gardenary, who makes it possible to reach the world with our message, our education, and our methods to get us closer each day to living in a world full of gardeners.

To Sarah Simon for her gorgeous artwork and Eric Kelley for the incredible cover photo, thank you for your artistry and incredible eye for garden beauty.

To those who inspired me to write a book about gardening habits: BJ Fogg, James Clear, and Charles Duhigg.

To Jason, my husband—I love you and the life we're growing together so very much. And to Carolyn, Brennan, Rebekah, and Elaine, thank you for all the moments of grace you've shown me as I worked on this book. And for inspiring me to write for a future that's worth growing into.

To my parents and sister, thank you for giving me a foundation and family that I've always felt secure and settled in. All that I've grown is a result of my deep roots.

And of course to the Gardenary community—wow, my friends. *You* are changing the world by growing in your corner of it. I'm so honored that you come to Gardenary for inspiration and education, and I count on you to make the gardening habit ordinary alongside me. I love you as much as I love my kitchen garden (and that's quite a bit!).

About
the Author

Nicole Johnsey Burke is the founder and owner of Gardenary, wife to Jason, and mom to four children. She loves working out, hiking, and going on adventures with her family and currently lives and gardens in Nashville, Tennessee.

She started Gardenary, Inc., in 2017 after her fast success with her first kitchen-garden-installation business. Today Gardenary makes gardening possible for everyone with online courses, books, and top-of-the-line garden supplies. Gardenary has taught more than 5,000 students through the Kitchen Garden Academy, trained more than 1,500 Gardenary-certified consultants, and supports thousands of gardeners around the world with Gardenary memberships.

Burke has written three books: *Kitchen Garden Revival* (2020), *Leaves, Roots & Fruit* (2023), and *The 5-Minute Gardener* (2025). She posts daily videos and tutorials to Gardenary's two million social followers. Her work has been featured by *Southern Living, This Old House, Modern Farmer,* and the Garden Club of America. You can follow her on social at **@heynicoleburke** and her companies at **@gardenaryco.**

Notes

_____

_____

_____

_____

_____

_____

_____

_____

_____

_____

_____

_____

_____

_____

Notes

_____

_____

_____

_____

_____

_____

_____

_____

_____

_____

_____

_____

_____

_____

Notes

Notes

Hay House Titles
of Related Interest

YOU CAN HEAL YOUR LIFE, the movie,
starring Louise Hay & Friends
(available as an online streaming video)
www.hayhouse.com/louise-movie

THE SHIFT, the movie,
starring Dr. Wayne W. Dyer
(available as an online streaming video)
www.hayhouse.com/the-shift-movie

***

*FOOD BABE FAMILY: More Than 100 Recipes and Foolproof
Strategies to Help Your Kids Fall in Love with Real Food:
A Cookbook,* by Vani Hari

*HEALING ADAPTOGENS: The Definitive Guide to Using Super
Herbs and Mushrooms for Your Body's Restoration, Defense, and
Performance,* by Tero Isokauppila and Danielle Ryan Broida

REAL SUPERFOODS: Everyday Ingredients to Elevate Your Health,
by Ocean Robbins and Nichole Dandrea-Russert, MS, RDN

*WILD REMEDIES: How to Forage Healing Foods and Craft Your
Own Herbal Medicine,* by Rosalee de la Foret & Emily Han

All of the above are available at your local bookstore,
or may be ordered by contacting Hay House (see next page).

***

We hope you enjoyed this Hay House book. If you'd like to receive our online catalog featuring additional information on Hay House books and products, or if you'd like to find out more about the Hay Foundation, please contact:

Hay House LLC, P.O. Box 5100, Carlsbad, CA 92018-5100
(760) 431-7695 or (800) 654-5126
www.hayhouse.com® • www.hayfoundation.org

———

Published in Australia by:
Hay House Australia Publishing Pty Ltd
18/36 Ralph St., Alexandria NSW 2015
Phone: +61 (02) 9669 4299
www.hayhouse.com.au

Published in the United Kingdom by:
Hay House UK Ltd
1st Floor, Crawford Corner,
91–93 Baker Street, London W1U 6QQ
Phone: +44 (0)20 3927 7290
www.hayhouse.co.uk

Published in India by:
Hay House Publishers (India) Pvt Ltd
Muskaan Complex, Plot No. 3,
B-2, Vasant Kunj, New Delhi 110 070
Phone: +91 11 41761620
www.hayhouse.co.in

———

Let Your Soul Grow

Experience life-changing transformation—one video at a time—with guidance from the world's leading experts.

www.healyourlifeplus.com

Free e-newsletters from Hay House, the Ultimate Resource for Inspiration

Be the first to know about Hay House's free downloads, special offers, giveaways, contests, and more!

 Get exclusive excerpts from our latest releases and videos from *Hay House Present Moments*.

 Our *Digital Products Newsletter* is the perfect way to stay up-to-date on our latest discounted eBooks, featured mobile apps, and Live Online and On Demand events.

 Learn with real benefits! *HayHouseU.com* is your source for the most innovative online courses from the world's leading personal growth experts. Be the first to know about new online courses and to receive exclusive discounts.

 Enjoy uplifting personal stories, how-to articles, and healing advice, along with videos and empowering quotes, within *Heal Your Life*.

Sign Up Now!

Get inspired, educate yourself, get a complimentary gift, and share the wisdom!

Visit www.hayhouse.com/newsletters to sign up today!

 HAY HOUSE

 HAY HOUSE online learning